A Guide to the Study of
ELECTRODE KINETICS

A Guide to the Study of
ELECTRODE KINETICS

H. R. THIRSK
J. A. HARRISON
School of Chemistry
The University
Newcastle-upon-Tyne
England

 1972

ACADEMIC PRESS · London and New York

ACADEMIC PRESS INC. (LONDON) LTD.
24/28 Oval Road,
London NW1

United States Edition published by
ACADEMIC PRESS INC.
111 Fifth Avenue
New York, New York 10003

Library of Congress Catalog Card Number: 74–172365
ISBN: 0–12–687750–5

PRINTED IN GREAT BRITAIN BY
ROYSTAN PRINTERS LIMITED
Spencer Court, 7 Chalcot Road
London NW1

Preface

The aim of this book is to present, in as brief a way as possible, the principles of the analysis of electrochemical measurements. It is assumed that the reader has a little experience of electrochemical techniques as outlined on pages 20 to 23. Some of the byways which delight the expert are missing, particularly the theory of electron transfer and the theory of the double layer, since, for the breadth of application of the book these difficult matters are not of primary relevance. Equally, the arithmetic behind the electrochemical techniques is not discussed in great detail, as this again is definitely a specialist area. What we have attempted is to cover a substantial number of electrochemical techniques and reaction schemes as are adequately worked out at the present moment. The actual schemes, apart from those of Chapter 3, are listed in Chapter 2, page 29. To keep the book to a reasonable size, no attempt has been made to illustrate each of the various treatments by actual examples, but adequate references are always made to appropriate electrochemical studies. By so doing, we felt that a clearer balance could be made between the various techniques when outlined in this comparatively uncomplicated manner. The treatment thus appears in a somewhat condensed form and is not without difficulties; these, however, are inherent in the study of reactions in which mass transfer effects must be considered. References, in the main, are to research monographs and the literature, although in the General References which follow this Preface, a number of text books are included which are considered to be particularly helpful. The research monograph has, to a large extent, replaced the traditional text book as a source of information, and it is from the former and the literature that the reader is recommended to put flesh on the bones.

Some of the material in this book is based on the M.Sc. course in electrochemistry which has been given at Newcastle for a number of years. In addition, one of us (H.R.T.) gave an invited course of four lectures on modern electrochemical kinetics at the 1968 May Meeting of the American Electrochemical Society, which was specifically intended as an introductory course. The lecturer gained a great deal from the associated informal discussions concerning the problems associated with a reasonably rapid mastery of electrochemical methods. This total experience has caused us to feel, most strongly, the need for a guide developed from these activities which would describe briefly the modern operational methods which are of routine use in our own and other laboratories; we therefore undertook a thorough rewriting of the above material as a shared task. The first three chapters (J.A.H.) describe methods by which the measurement of electrical parameters gives kinetic information. The final chapter gives an account of non-electrical and *in situ* methods of detecting intermediates, and examining electrode surfaces and products (H.R.T.). We hope the whole will be useful to postgraduate students and research workers alike.

In conclusion, we would like to thank Mrs. E. Lewis of the Graphics Section, Department of Photography, University of Newcastle-upon-Tyne, who so ably drew the figures in the book, and also our colleagues and students over many years for innumerable discussions on electrochemical topics.

University of Newcastle-upon-Tyne   J. A. Harrison
January, 1972   H. R. Thirsk

Contents

General References

TEXT BOOKS

Damaskin, B. B., "Principles of Current Methods for the Study of Electrochemical Reactions". McGraw-Hill, New York (1967).

Vetter, K. J., "Electrochemical Kinetics". Academic Press, London and New York (1967).

Delahay, P., "Double Layer and Electrode Kinetics", Interscience, New York (1965).

Delahay, P., "New Instrumental Methods in Electrochemistry", Interscience, New York (1954).

"Physical Chemistry", (Eds., Eyring, H., Henderson, D., Jost, W.), Vols, IXA, IXB "Electrochemistry", Academic Press, London and New York (1970).

Levich, V. G., "Physico-Chemical Hydrodynamics", Prentice-Hall, New York (1962).

REVIEW SERIES

"Modern Aspects of Electrochemistry" (Eds., Bockris, J. O'M., Conway, B. E.), Plenum Press, New York.

"Advances in Electrochemistry and Electrochemical Engineering", (Eds., Delahay, P., Tobias, C. W.), Interscience, New York.

"Electroanalytical Chemistry", (Ed., Bard, A. J.,) Dekker, New York

Regular reviews which appear in the American journal *Analytical Chemistry*.

List of Symbols

There are certain symbols of limited and special use which are only defined as used in the text:

A area

A rate of nucleation in Chapter 3. Defined in two ways by eq. (3.5) and (3.8)

$\sum A_n$ function of the extent of coverage and r_2

b $\alpha n f v$

C_O^s, C_R^s surface concentrations of species O and R (at $x = 0$)

C_O^b, C_R^b bulk concentrations of species O and R

$(C_O^b)_i$ initial bulk concentration

$\bar{C}$ Laplace transform of the concentration

D diffusion coefficient. Also with appropriate subscripts identifying species

E^0 standard potential

E potential of working electrode

$E_{\frac{1}{2}}$ half wave potential

E_e potential at which $i = 0$

E_i initial potential

E_p peak potential

f F/RT

F Faraday

$F(\alpha)$ structural parameter, eq. (2.235)

h height of a nucleus

i current in amps. Chapters 1 and 2; in Chapter 3, current density

i_d diffusion limited current

i_0 exchange current density, amps. cm^{-2}

i_l limiting current

i_p peak current, eq. (2.132)

i_p' peak current, eq. (2.139)

$\left.\begin{array}{l} \bar{i} \\[6pt] \bar{i}(s) \end{array}\right\}$ Laplace transform of the current

I current density due to the interfacial reaction; that is the current after correction for diffusion effects, amps. cm^{-2}

j $\sqrt{-1}$

k_f a forward rate constant

k_b a backward rate constant

k_1, k_{-1} potential independent characteristic forward and backward rate constants for cathodic reactions

k_1', k_{-1}' potential independent forward and backward rate constants

k_{sh} forward rate on the surface

k_1, k_2 Chemical rate constants are variously k_1 and k_2 as defined in the original papers. They are also defined in the reaction schemes immediately before use

K chemical equilibrium constant

K_d σ, see eq. (2.166)

l half distance between step lines in adatom model

M molecular weight

n number of electrons transferred

n_D number of electrons in reaction at disc electrode

n_R number of electrons in reaction at ring electrode

N number of nuclei, see eq. (3.5)

N collection efficiency; ring disc electrode eq. (2.230)

N_0 initial number of nuclei

r radius of nucleus

r_1 radius of disc electrode

r_2 radius from the centre of a disc electrode to the outer edge of the isolating ring

r_3 radius from the centre of the disc to the outer edge of the metal ring electrode

         For general reference for r_1, r_2 and r_3 see Fig. 2.37

r_2' radius of diffusion cylinder surrounding an active site

R gas constant

s Laplace variable; see definition of the Laplace transform eq. (2.7)

S area as defined by eq. (3.1) and Fig. 3.1

S_1 area covered per unit area when overlap is taken into account

S_{1ex} area covered, per unit area of electrode, by isolated nuclei

t time

T absolute temperature

u time for growth of nuclei

v potential sweep rate

V volume

Z impedance

$Z(s)$ eq. (2.158)

$Z(j\omega)$ eq. (2.159)

$\bar{Z}(s)$ transform of impedance

α transfer coefficient, E negative going as i, positive for a cathodic reaction, increases

β transfer coefficient, E negative going as i, positive for an anodic reaction, increases

δ Nernst diffusion layer thickness

η $(E - E_e)$

η' $\eta - (E - E^0)$

θ coverage; also with subscripts identifying species

θ $R_D = \dfrac{1}{Anfi_0}$ a.c. theory; eq. (2.164)

θ ring disc theory; function defined in eq. (2.228)

λ $\dfrac{(k_f + k_b)}{v}\dfrac{RT}{nF}$

λ_d $\dfrac{k_d C_A{}^B}{v}\dfrac{RT}{nF}$ eq. (2.155)

λ_m $\dfrac{kC_2}{\alpha na}\dfrac{RT}{nF}$ eq. (2.145)

v kinematic viscosity, eq. (2.195)

ρ density

σ $K_d = \dfrac{1}{An^2 Ff2^{\frac{1}{2}}}\left[\dfrac{1}{C_0{}^b D_0{}^{\frac{1}{2}}} + \dfrac{1}{C_R{}^b D_R{}^{\frac{1}{2}}}\right]$ eq. (2.166)

τ transition time

ϕ_1 potential at the inner Helmholtz plane

ϕ_2 potential at the outer Helmholtz plane

ϕ_m potential at the electrode surface

ϕ_s potential in the solution taken as zero

ϕ phase angle in a.c. method, calculated by eq. (2.167)

Ψ current function defined as $\dfrac{i}{nFAD\,C\,(nFv/RT)}$, eq. (2.138)

ω frequency; radians $\sec^{-1}$

ω_R rotation speed of rotating disc

Chapter 1

Stationary Methods

INTRODUCTION

FACTORS WHICH DETERMINE CURRENT–VOLTAGE CURVES

In this chapter some of the ways of analysing current–voltage curves will be discussed. In order to obtain a stationary current voltage curve at a single electrode it is necessary to have some method of controlling the diffusion layer thickness, such as polarography or the rotating disc. The specific techniques are discussed in Chapter 2.

MEASUREMENT AT A SINGLE ELECTRODE

An electrolytic cell is composed of two electrodes. However in kinetic investigations only one electrode is investigated. It is usual to isolate the effect of one electrode from the other by controlling it's potential. The most satisfactory way of doing this is with a potentiostat and Luggin capillary. The practical details are discussed at the end of this chapter.

CURRENT–VOLTAGE BEHAVIOUR

The effect of applying potential to a single electrode, M, in an electrolyte, HCl say, is shown diagrammatically in Fig. 1.1. Thermodynamic data can often be used to decide, in principle, which reactions are possible. Whether

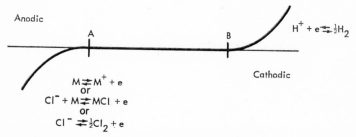

FIG. 1.1. Form of the current voltage curve for a single metal electrode M in an electrolyte M, HCl.

1

a particular reaction goes is a matter of experiment. As shown in the diagram, current only flows more cathodic than B or anodic than A. If a more reducible substance than H^+ is present, then current is observed as a wave prior to hydrogen evolution.

DOUBLE LAYER

In the region AB (Fig. 1.1) no permanent current flows and the electrode is said to be perfectly polarisable. Change of potential in this region changes the charge on both sides of the interface, as in a parallel plate condenser. A great deal is now known about the structure of the double layer. The main features are shown in Fig. 1.2. Some anions penetrate to the inner Helmholtz plane and are specifically adsorbed. Cations and unadsorbed anions only approach as far as the outer Helmholtz plane. A layer of water exists between the inner Helmholtz plane and the metal surface. The quantities which are measured to deduce double layer structure are differential capacity, interfacial tension, surface potential. The reader is referred to specialist reviews for further details of the interpretation of the effect of inorganic ions[1,2,3,4,5] and organic[6] ions at the mercury electrolyte interface.

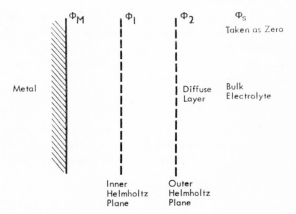

FIG. 1.2. Properties of the interface inferred from measurements in the double layer region AB of Fig. (1.1).

Although the interface can be investigated in the double layer region (at mercury) it is undoubtably also present when a reaction proceeds. Attempts have been made to match double layer properties measured in the absence of reactant, with the reaction kinetics.[7(a,b),8] An advance has been made in this direction in that the double layer capacity can now be measured in the presence of a reaction (see Sluyters a.c. method in Chapter 2).[9] Some reactions change the expected double layer capacity at mercury and the reactant ion is itself adsorbed Tl^+, Hg_2^{2+}, K^+, Pb^{2+}), others do not (Zn^{2+}). Measurements of this kind can give more detailed electrochemical information although little has yet been carried out in this direction.

Differential capacity measurements have almost exclusively been measured on mercury. With a deeper understanding of roughness and surface structure,[10] it will be possible to obtain meaningful results on other metals. An attempt along these lines has been made with silver,[11]

In general, however, in electrode kinetics the details of the interface have, as yet not proved to be of overriding importance. The presence of excess inert electrolyte, to minimise migration, ensures that the potential at the reaction site remains constant as the overall potential of the electrode is changed. The face that many reactions under these conditions have a straight Tafel plot of the correct slope and also have the correct dependence of current on reactant concentration at fixed potential (with respect to the reference electrode) supports this argument. If this is not the case experimentally then the structure of the double layer must be taken into account.

DEFINITION OF INTERFACIAL RATES OF REACTION

The basic electrochemical process which occurs when an inert metal is in contact with a redox process is

$$O + ne \underset{k_b}{\overset{k_f}{\rightleftharpoons}} R. \tag{1.1}$$

The current which flows as in equation (1.1) at a particular potential, i.e. the Faradaic current, can be expressed as the difference between the forward and reverse rates, as in chemical kinetics

$$i = nFA(k_f C_O^s - k_b C_R^s) \tag{1.2}$$

k_f, k_b are constants at fixed potential. C_O^s, C_R^s are the concentrations of O and R at the surface, where they are discharged. n is the number of electrons involved, A is the electrode area and F is the Faraday. The activity coefficients, which are usually determined by the environment, i.e. the base electrolyte, are included in k_f and k_b. Equivalent to equation (1.2) is the statement that the overall current is given by the partial currents

$$i = i_f - i_b. \tag{1.3}$$

The overall current i will be considered positive for a cathodic reaction on which equation (1.1) goes from left to right. It is however convenient if one is dealing with an anodic reaction to let an anodic current be positive (see below for the definition of potential in this case). k_f and k_b are expected to be exponential functions of potential. This assumption can be justified by transition state theory. Many accounts are available and it is not necessary to repeat them here.[12,13] The main property which determines the current is the potential E of the working electrode with respect to a reference electrode. k_f, k_b are usually defined in terms of E as follows,

$$k_f = k_1 \exp{(-\alpha n f E)} \tag{1.4}$$

$$k_b = k_{-1} \exp{((1 - \alpha)nfE)} \tag{1.5}$$

where $f = F/RT$. E is considered to be negative going as i, which is positive, increases, In some places in the text, in order to match current literature, it will be convenient to reverse the convention and to let an anodic current have a positive sign and consider E to be negative going as i increases. The constant will, in this case, be called β.

Equations (1.4) and (1.5) can now become

$$k_f = k_1' \exp{(-\beta n f E)} \tag{1.6}$$

$$k_b = k_{-1}' \exp{((1 - \beta)nfE)} \tag{1.7}$$

for the reaction

$$R \underset{k_b}{\overset{k_f}{\rightleftharpoons}} O + ne \tag{1.8}$$

$k_1, k_{-1}, k_1', k_{-1}'$ are potential independent constants, and α, β are empirical constants with a value approximately 0·5. Attempts have been made to give α, β theoretical significance but, as yet, no completely satisfactory theory has emerged.[14,5] In the rest of this chapter equations (1.4), (1.5) for the cathodic reaction will be discussed. Exactly the same equations will apply to the anodic reaction. The numerical values of k_1, k_{-1} depend on the reference electrode system, that is $k_f = k_1$, $k_b = k_{-1}$, at $E = 0$. In order to define constants equivalent to k_1, k_{-1} which reflect the nature of the electrochemical process itself, it is usual to use two alternatives. The first is to define k_f, k_b with respect to E^0, the standard potential.

$$k_f = k_{sh} \exp{\left(-(E - E^0)\alpha n f\right)} \tag{1.9}$$

$$k_b = k_{sh} \exp{((E - E^0)(1 - \alpha)nf)}. \tag{1.10}$$

The second method is to define k_f, k_b with respect to the potential E_e, at which experimentally $i = 0$. Using the definition for overpotential

$$\eta = E - E_e \tag{1.11}$$

$$k_f = \frac{i_0}{nFC_O{}^b} \exp\left(-\alpha nf\eta\right) \tag{1.12}$$

$$k_b = \frac{i_0}{nFC_R{}^b} \exp\left[(1-\alpha)nf\eta\right] \tag{1.13}$$

i_0 is the exchange current.

E^0 is useful as a reference point when only one member of the redox couple is present and E_e can not easily be determined. It is clear that very often experiments will be carried out by the electrolysis of just one substance and E_e cannot be measured. Substituting equations (1.12), (1.13) in equation (1.2) gives

$$\frac{i}{A} = i_0 \left[\frac{C_O{}^s}{C_O{}^b} \exp\left(-\alpha nf\eta\right) - \frac{C_R{}^s}{C_R{}^b} \exp\left[(1-\alpha)nf\eta\right] \right]. \tag{1.14}$$

STATIONARY METHODS

Equation (1.14) describes quite generally the relation between overall current and potential when the concentration of the reacting species is different from the bulk concentration. This situation arises because as the electrochemical reaction procedes new reactant must arrive at the interface by diffusion. It will be shown in Chapter 2 that the current to a fixed planar electrode, in a still solution, has no stationary state. The question is under what conditions can equation (1.14) be observed and used. If the diffusion layer is fixed then a stationary current for a given potential can be observed. Techniques which fulfill this condition are polarography, rotating disc, thin layer cell, spherical micro-electrode; the last mentioned is rarely used. An alternative which is often used in research is to construct current–voltage from say current–time curves at fixed time.

Treatment of diffusion

If a diffusion gradient exists between the surface and the bulk of solution, the movement of ions in a concentration gradient is controlled by Fick's first and second laws (this topic is continued in Chapter 2).

$$\text{flux} = D\left(\frac{\partial C}{\partial x}\right) \quad \text{Fick's first law} \tag{1.15}$$

$$\left(\frac{\partial C}{\partial t}\right) = D\left(\frac{\partial^2 C}{\partial x^2}\right) \quad \text{Fick's second law.} \tag{1.16}$$

DIFFUSION LAYER MODEL

If Fick's second law is solved by integration for a stationary situation, then the flux per unit area is a constant given by

$$\text{flux} = \frac{D(C^b - C^s)}{\delta} \tag{1.17}$$

where δ is the diffusion layer at which the concentration is maintained at $C_O{}^b$ (see rotating disc and polarography).

SIMPLE ELECTRON TRANSFER

LIMITING CASES OF EQUATION (1.14)

Two limiting cases arise immediately from equation (1.14). Where i_0 is small then the bulk concentration is little disturbed at the interface.

$$C_O{}^s = C_O{}^b, \; C_R{}^s = C_R{}^b \tag{1.18}$$

then

$$\frac{i}{A} = i_0(\exp(-\alpha nf\eta) - \exp\{(1-\alpha)nf\eta\}). \tag{1.19}$$

Fig. 1.4 shows equation (1.19) as a function of α. Rearranging the equation

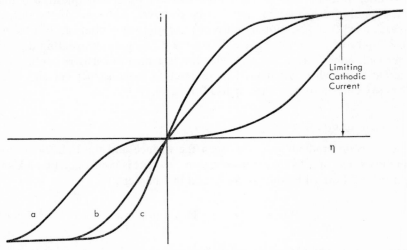

FIG. 1.3. Form of a current–voltage curve according to equation (1.14). It is assumed that the curve is measured at a fixed diffusion layer, i.e. directly by polarography, rotating disc, or indirectly from the measurement of transients.

(1.19) gives

$$\ln \frac{i}{A[1 - \exp nf\eta]} = \ln i_0 - \alpha nf\eta. \qquad (1.20)$$

The exponential separate anodic and cathodic branches become combined into a linear plot through $\eta = 0$. The slope is αnf and the intercept, dashed in the Fig. 1.5, is $\ln i_0$. This is the basis of comparing experiment and theory (it must be emphasised that diffusion has already been removed) in order to calculate i_0.

At high potentials or if the reaction is intrinsically irreversible then equation (1.19) becomes

$$\ln \frac{i}{A} - \ln i_0 = -\alpha nf\eta \qquad (1.21)$$

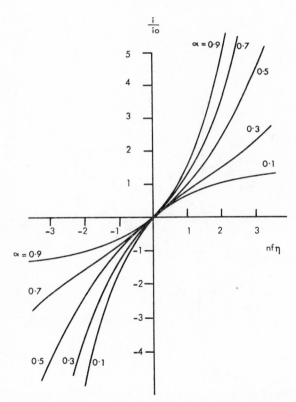

FIG. 1.4. Current–voltage curve according to equation (1.19). This is observed in practice when diffusion has been extrapolated out (most straight forwardly by means of the rotating disc) or i_0 is low.[32]

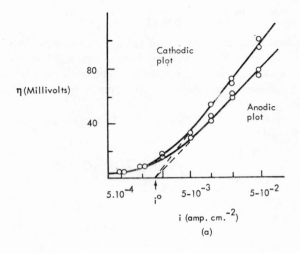

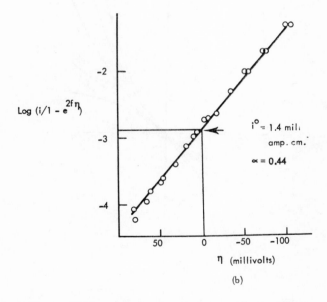

FIG. 1.5. Plot of equation (1.19) and the corresponding plot of equation (1.20).

which is the well known Tafel equation of slope αnf

$$\frac{\mathrm{d}\ln i}{\mathrm{d}\eta} = -\alpha nf. \tag{1.22}$$

In particular expanding the exponentials in equation (1.19) about $\eta = 0$ shows that

$$\frac{\mathrm{d}i}{\mathrm{d}\eta} = -\frac{i_0}{nf} \tag{1.23}$$

these equations are used extensively. On the other hand, when i_0 is large then the two terms in the bracket in equation (1.14) will be nearly equal and only small differences between them will be necessary to generate a large current. If

$$i_0 \frac{C_{\mathrm{O}}^{s}}{C_{\mathrm{O}}^{b}} \exp\left(-\alpha nf\eta\right) = i_0 \frac{C_{\mathrm{R}}^{s}}{C_{\mathrm{R}}^{b}} \exp\left\{(1-\alpha)nf\eta\right\} \tag{1.24}$$

then

$$\frac{C_{\mathrm{O}}^{s}}{C_{\mathrm{O}}^{b}} \frac{C_{\mathrm{R}}^{b}}{C_{\mathrm{R}}^{s}} = -\exp\left(nf\eta\right). \tag{1.25}$$

Equation (1.25) is equivalent to

$$\eta = \frac{1}{nf} \ln \frac{C_{\mathrm{O}}^{s}}{C_{\mathrm{R}}^{s}} + \frac{1}{nf} \ln \frac{C_{\mathrm{R}}^{b}}{C_{\mathrm{O}}^{b}}. \tag{1.26}$$

Given the Nernst equation

$$E_e = E^0 + \frac{1}{nf} \ln \frac{C_{\mathrm{O}}^{b}}{C_{\mathrm{R}}^{b}} \tag{1.27}$$

then equation (1.26) becomes

$$E = E^0 + \frac{1}{nf} \ln \frac{C_{\mathrm{O}}^{s}}{C_{\mathrm{R}}^{s}}. \tag{1.28}$$

Equation (1.28) is the Nernst equation at the electrode surface. It is assumed, as before, that the individual activity coefficients, which are determined by the total ionic strength (see Debye–Hückel equation) cancel. The kinetic factor i_0 has disappeared from equation (1.26) and the reaction is driven at constant potential only by the concentration gradient and the fixed ratio $C_{\mathrm{O}}^{s}/C_{\mathrm{R}}^{s}$. The concentration gradient is given by Fick's first law of diffusion, and is linear in the stationary state. As material cannot accumulate at $x = 0$ then the rate of electrochemical reaction must balance the rate of diffusion. Given a fixed diffusion layer δ and the result of the last section

$$i = \frac{nFAD_O(C_O^b - C_O^s)}{\delta} = - \frac{nFD_R(C_R^b - C_R^s)}{\delta} . \tag{1.29}$$

The reader is referred to Chapter 2 for a detailed discussion of the diffusion layer in connection with the rotating disc and polarography.

Definition of reversibility

If a reaction follows equation (1.14) or (1.19), it is said by electrochemists to be reversible. If it follows equations (1.24) and (1.25) it is said to be perfectly reversible. A reaction which only goes in one direction, that is has a rate given by

$$i = nFAk_f C_O^s \tag{1.30}$$

where k_f is again defined by equation (1.9) or equation (1.12), is known as a completely irreversible reaction. This condition arises either if a large potential is applied to the electrode so that the back reaction is negligible or if the reaction is intrinsically irreversible. Electrochemists usually wish to know if a reaction is completely reversible, reversible or irreversible. The rules for deciding are given in Chapter2.

GENERAL TREATMENT OF EQUATION (1.14)

The general expression of a moderately reversible reaction in which both components are present in the bulk solution is given by equation (1.14). The concentrations at the surface are given by equation (1.29).

$$\frac{C_O^s}{C_O^b} = \frac{(i_l)_O - i}{(i_l)_O} \tag{1.31}$$

$$\frac{C_R^s}{C_R^b} = \frac{i + (i_l)_R}{(i_l)_R} \tag{1.32}$$

where the diffusion currents $(i_l)_O$ and $(i_l)_R$ corresponding to the species O, R are given a positive sign. Substituting the concentrations given by equations (1.31), (1.32) into equation (1.14) gives

$$i = nFAi_0 \left\{ \frac{(i_l)_O - i}{(i_l)_O} \exp\left(-\alpha nf\eta\right) - \frac{i + (i_l)_R}{(i_l)_R} \exp\left\{(1 - \alpha)nf\eta\right\} \right\} \tag{1.33}$$

and a simple rearrangement[15] gives

$$\ln\left\{\left(\frac{1}{i} + \frac{1}{(i_l)_O}\right) - \left(\frac{1}{(i_l)_R} + \frac{1}{i}\right)\exp(nf\eta)\right\} = \ln\frac{1}{(i_0 nFA)} + \alpha nf\eta. \quad (1.34)$$

In the general case, a plot of the L.H.S. against η allows estimation of i_0 as the intercept and α as the slope.

However it is more convenient and often essential to observe a current voltage curve with only one component present in the bulk. It is now easier to use E^0 as the reference point for potential as E_e cannot be easily experimentally determined. Using the definitions of equations (1.9), (1.10) then

$$i = nFA(k_f C_O{}^s - k_b C_R{}^s) \quad (1.35)$$

becomes

$$i = nFAk_{sh}\{C_O{}^s\exp[-\alpha(E-E^0)nf] - C_R{}^s\exp[(1-\alpha)(E-E^0)nf]\}. \quad (1.36)$$

The flux does not accumulate hence for one component in the bulk, O say, equations (1.31), (1.32) are

$$C_O{}^b - C_O{}^s = \frac{D_R}{D_O}C_R{}^s \quad (1.37)$$

and

$$\frac{C_O{}^b - C_O{}^s}{C_O{}^b} = \frac{i}{(i_l)_O} \quad (1.38)$$

so substituting for $C_O{}^s$, $C_R{}^s$ in equation (1.36) gives

$$i = nFAk_{sh}\left\{\left(\frac{(i_l)_O - i}{(i_l)_O}\right)C_O{}^b\exp[-\alpha f(E - E^0)] - \frac{D_O}{D_R}C_O{}^b\frac{i}{i_l}\right.$$

$$\left. \times \exp[(1-\alpha)(E - E^0)nf]\right\} \quad (1.39)$$

and the general equation becomes

$$\ln\left\{\left(\frac{1}{i} - \frac{1}{(i_l)_O}\right)C_O{}^b - \frac{D_O}{D_R}\frac{C_O{}^b}{(i_l)_O}\exp[(E - E^0)nf]\right\} = \ln\frac{1}{nFAk_{sh}}$$

$$+ \alpha nf(E - E^0) \quad (1.40)$$

which once again could be calculated from experimental data by plotting the L.H.S. against $E - E^0$, k_{sh} can be converted to i_0 by equation (1.47).

Completely irreversible reactions are much easier to interpret. If a reaction takes place away from E^0 then the back reaction can be ignored and equation (1.30) for one component becomes

$$i = nFAk_{sh} \left(\frac{(i_l)_O - i}{(i_l)_O} \right) C_O{}^b \exp\left(- \alpha f(E - E^0) \right) \qquad (1.41)$$

hence

$$\ln \frac{(i_l)_O - i}{i} + \ln \frac{C_O{}^b nFAk_{sh}}{(i_l)_O} = \alpha f(E - E^0). \qquad (1.42)$$

A similar expression clearly holds if E^0 is unknown and definition (1.4) is used

$$\ln \frac{(i_l)_O - i}{i} + \ln \frac{C_O{}^b nFAk_1}{(i_l)_O} = \alpha f E. \qquad (1.43)$$

A plot of $\ln [(i_l)_O - i]/i$ against E is very well known in electrochemistry is very often referred to as correcting for concentration overpotential.

The situation for a perfectly reversible reaction is straightforward. The general case when both components are present in the bulk gives, by equating the two RHS terms in equation (1.14),

$$\ln \frac{(i_l)_O - i}{i + (i_l)_R} + \ln \frac{(i_l)_R}{(i_l)_O} = nf\eta \qquad (1.44)$$

and this can easily be plotted to give only the value of n, i_0 disappears as the reaction is fast and totally controlled by concentration gradients. When one component only is present the equation (1.44) becomes

$$\ln \frac{(i_l)_O - i}{i} + \ln \frac{D_R}{D_O} = nf(E - E^0). \qquad (1.45)$$

Concentration dependence

There are two ways of determining the concentration dependence. i_0 and its concentrations dependence can be measured, or that of the current at fixed electrode potential (against an external reference electrode). It is preferable to measure i_0, as this is independent of junction potential. However i_0 is rarely accessible as the electrochemical step must be reversible and not too fast. i_0 and k_{sh} are interrelated. Introducing the Nernst equation

$$E_e = E^0 + \frac{1}{nf} \ln \frac{C_O{}^b}{C_R{}^b} \qquad (1.46)$$

into equations (1.9) and (1.12) gives

$$i_0 = k_{sh}(C_O{}^b)^{1-\alpha}(C_R{}^b)^\alpha \qquad (1.47)$$

i_0 has the advantage, as a characteristic measure, that it is a directly accessible quantity if an equilibrium potential at $i = 0$ is experimentally measurable. The concentration dependence of i_0 from equation (1.47)

$$\left(\frac{d \ln i_0}{d \ln C_O{}^b}\right)_{C_R{}^b} = 1 - \alpha \qquad (1.48)$$

$$\left(\frac{d \ln i_0}{d \ln C_R{}^b}\right)_{C_O{}^b} = \alpha \qquad (1.49)$$

is a diagnostic test. k_{sh} is concentration independent but implies a knowledge of α.

Similar arguments can be used for more complicated reaction schemes. To consider a well known example[17] where a complex is involved in the electrochemical step

$$MX_v^{(v-1)-} + e \rightleftharpoons M + vX^- \qquad (1.50)$$

$$i_0 = k_{sh}[MX_v]^{1-\alpha}[X^-]^{v\alpha}[M]^\alpha \qquad (1.51)$$

and

$$\frac{d \ln i_0}{d \ln X^-} = v + \frac{(1-\alpha)F}{RT}\left(\frac{d \ln E_e}{d \ln X^-}\right) \qquad (1.52)$$

hence v can be found from experimental quantities. In general i_0 would be measured as a function of all the possible concentrations.

A simple first order electrochemical step in O should plot according to equations (1.45) or (1.43). The plot of $\ln\{[(i_l)_O - i]/i\}$ against E should be independent of concentration. If it is inconvenient to use this method, if i_l is inaccessible for example, then a simpler more general method can be attempted. Consider that diffusion is removed (either by an extrapolation of the type shown in Chapter 2, or if i_0 is small). Equation (1.2) can be generalised to more than one reactant. Assuming that equation (1.50) operates

$$i = nFA(k_f C_{MX_v^-} - k_b(C_{X^-})^v) \qquad (1.53)$$

According to equation (1.3) at constant E

$$i_f = nFAk_f C_{MX_v} \qquad (1.54)$$

$$i_b = nFAk_b(X^-)^v. \qquad (1.55)$$

As

$$i_f = \frac{i}{1 - \exp{(nf\eta)}} .$$ (1.56)

The i_f or i_b can be plotted against concentration logarithmically to find the order of reaction. If η is not known or if the reaction is irreversible equations (1.54), (1.55) can be used directly at sufficiently large E.

Influence of double layer structure

Until now it has been assumed that the potential is uniform up to the interface. However if the reacting ion sees a different potential at the site where it is discharged for example at the outer Helmholtz layer, with potential ϕ_2, then more precisely

$$i_t^0 = k_{sh}^{t} (C_O^b)^{1-\alpha} (C_R^b)^{\alpha}$$ (1.57)

where the true parameters have superscript t, and the apparent value can be corrected if ϕ_2 is known. ϕ_2 is normally controlled by the indifferent electrolyte and can be measured in the absence of the reacting ion. In many cases ϕ_2 is, however, small. This situation where the reacting ion is adsorbed and experiences the potential of the inner layer, ϕ_1 has been discussed by Parsons.[18(a,b)]

EXPANSION OF EQUATION (1.14)

A further property of equation (1.14) needs to be considered. Suppose that equation (1.14) is expanded for small values of η about $\eta = 0$ then

$$\frac{i}{A} = i_0 \left[\frac{\eta}{f} + \frac{C_R^s}{C_R^b} - \frac{C_O^s}{C_O^b} \right] .$$ (1.58)

Equation (1.58) is necessary later in Chapter 2 in calculating the theory for the galvanostatic and a.c. methods.

CONSECUTIVE ELECTRON TRANSFER

Consider the more complex scheme

$$O + e \rightleftharpoons R_1$$ (1.59)

$$R_1 + e \rightleftharpoons R_2.$$ (1.60)

This implies that an equilibrium[19]

$$O + R_2 \overset{K}{\rightleftharpoons} 2R_1$$ (1.61)

and possibly

$$2R_1 \overset{K_3}{\rightleftharpoons} D$$ (1.62)

exist in the solution away from the electrode. Only if R_1 has an appreciable lifetime will two distinguishable current voltage curves be observeable. The current voltage curves can be theoretically calculated in a number of cases.

Equation (1.59) reversible, equation (1.60) irreversible

The solution equilibrium (1.61), (1.62) cannot operate under these conditions and the calculation is straight forward. For a Nernst diffusion layer the flux due to equation (1.59)

$$i_1 = \frac{(i_1)_l}{C_O^b} (C_O^b - C_O^s) \tag{1.63}$$

is equal to the flux due to equation (1.60)

$$i_1 = AF k_{sh} C_{R_1}^s \exp[-\alpha f(E - E^0)] \tag{1.64}$$

which with the Nernst equation becomes, if only O is present in solution,

$$i = AF k_{sh} C_O^s \exp[-(1 + \alpha)f(E - E^0)]. \tag{1.65}$$

Elimination of C_O^s and use of $i = 2i_1$, allows i to be calculated. In more conventional form

$$\ln \frac{i}{2(i_1)_l - i} = \ln \frac{AF k_{sh} C_O^b}{(i_1)_l} - (1 + \alpha)(E - E^0)f. \tag{1.66}$$

This expression is characterised by a 40 mV Tafel slope given by $[(1 + \alpha)f/2.303$. When $i < (i_1)_l$ a plot of $\log i/E$ would have this slope directly.

Equation (1.59) irreversible, equation (1.60) irreversible

When both reactions are slow with different i_0 and α the predicted current–voltage relation is[20]

$$\frac{i}{A} = \frac{2(i_0)_1 \exp(-\alpha_1 f\eta)[1 - \exp - \alpha f\eta]}{1 + [(i_0)_1/(i_0)_2]\exp[(1 + \alpha_2 - 2_1)f\eta]}. \tag{1.67}$$

Equation (1.59) irreversible, equation (1.60) reversible

In this case the second reaction amplifies the current due to the first reaction

$$i = 2i_1 = 2FA k_{sh} \exp[(E - E^0)\alpha f]. \tag{1.68}$$

The Tafel slope is 120 mV.

Equation (1.59) *reversible, equation* (1.60) *reversible*

If reaction (1.61), (1.62) is kinetically hindered then the reaction will behave as a reversible wave with $2e$ overall. The shape will be equation (1.44) or equation (1.45). However when equilibrium (1.61) is taken into account[19] ($K_3 = 0$) the current voltage curve becomes

$$\ln \frac{2(i_l)_O - i(1 + \exp 2f\eta')}{i - (i_l)_O} = \ln p + f\eta' \tag{1.69}$$

when $C_{R_1}{}^b = C_{R_2}{}^b = 0$. p is given by $\sim K^{1/2}$. The general case when both O, R, S exist in the bulk and $K_3 \neq 0$ are discussed in detail in Ref. (14). Where $\eta' = \eta - (E_{1/2} - E^0)$.

CHEMICAL PROCESSES PRECEDING SINGLE ELECTRON TRANSFER

If the electroactive species is supplied to the interface by a chemical reaction as well as diffusion

$$Y \underset{k_2}{\overset{k_1}{\rightleftharpoons}} O$$

$$O + ne \rightleftharpoons R \tag{1.70}$$

the current–voltage curve can be easily calculated in one simple case. If the concentration of O is small so that the preceding equilibrium is to the left the transport of O to the surface can be described reasonably accurately by the reaction layer model.

REACTION LAYER MODEL

By analogy with the supply of O by diffusion, equation (1.17)

$$\text{flux} = \frac{D_O(C_O{}^b - C_O{}^s)}{\mu} \tag{1.71}$$

where μ is the reaction layer thickness. The limiting current is

$$i_l = \frac{D_O C_O{}^b}{\mu} nFA \tag{1.72}$$

but

$$\frac{C_y{}^b}{C_O{}^b} = K \tag{1.73}$$

and it can be shown that

$$\mu = \sqrt{D_O/k_2} \tag{1.74}$$

hence

$$i_l = nFAD_O^{1/2} C_Y{}^b \frac{k_1^{1/2}}{K^{1/2}}. \tag{1.75}$$

It is also quite simple to couple this with the diffusion layer model to calculate i_1 if the diffusion of C_Y is involved.

Current voltage curve

If the limiting current is independent of δ that is of stirring and $i_l < i_d$ then a current voltage curve can be predicted exactly as before equations (1.43), (1.45) in which i_l refers to the current controlled by the reaction layer.

CHEMICAL REACTIONS FOLLOWING THE ELECTRON TRANSFER

It becomes less feasible to analyse the current voltage curve in general. However two single cases have been given in the literature. In both cases the electron transfer is reversible.

ELECTRON TRANSFER WITH FIRST ORDER SUCCEEDING REACTION

The shape of the current voltage curve[16] for

$$O + ne \rightleftharpoons R \tag{1.76}$$

$$R \xrightarrow{k} B$$

in which the electron exchange step is reversible, is given by

$$\ln \frac{i_l - i}{i} + \ln \frac{k^{1/2} A D_R^{1/2} nF}{i_l} = \frac{nF}{RT}(E - E^0). \tag{1.77}$$

ELECTRON TRANSFER WITH SECOND ORDER SUCCEEDING REACTION

It can also be easily shown[16] that for

$$O + ne \rightleftharpoons R \tag{1.78}$$

$$2R \xrightarrow{k_d} B$$

In which the electron transfer is reversible

$$\ln \frac{i^{2/3}}{i_l - i} - \ln \frac{i_l}{(nFA)^{2/3}(D_R k_d)^{1/3}} = \frac{nF}{RT}(E^0 - E). \tag{1.79}$$

SURFACE PROCESSES

(i) Simple reactions of the type

$$A_{ads} \underset{k_b}{\overset{k_f}{\rightleftharpoons}} B_{ads} + e \tag{1.80}$$

are expected to follow the same exponential law as equation (1.14).

(ii) Scheme (1.80) can be extended to[39]

$$A \underset{k'_{-A}}{\overset{k'_A}{\rightleftharpoons}} A_{ads} \tag{1.81}$$

$$X + A_{ads} \underset{k_{-A}}{\overset{k_A}{\rightleftharpoons}} B_{ads} + e \tag{1.82}$$

$$X + B_{ads} \underset{k_f}{\rightarrow} C_{ads} + e \tag{1.83}$$

The reaction sequence is written for an anodic process. θ is the coverage with A or B.

Langmuir isotherm

Equilibrium for the first two reactions gives

$$k_A'C_A^b(1 - \theta_T) = k'_{-A}\theta_A \tag{1.84}$$

and

$$k_A\theta_A C_x^b = k_{-A}\theta_B. \tag{1.85}$$

θ_T is the total coverage $\theta_A + \theta_B$. C_x^b is assumed constant up to the interface. The rate determining step, assuming as always that bulk diffusion has been extrapolated out,

$$i = nFA\theta_B k_f C_x^b \tag{1.86}$$

k_f has now different units (sec^{-1}), if C_x^b is mols cm^{-2}. Equations (1.85), (1.86) give

$$\theta_B = \frac{k_A k_A'}{k_A k'_{A-}} C_A^b(1 - \theta_T) \tag{1.87}$$

hence

$$i = nFA \frac{k_A k_A'}{k_{-A} k'_{-A}} C_A^b(C_x^b)^2(1 - \theta_T)k_f \tag{1.88}$$

Definition (1.6) for k_f

$$k_f = k_1 \exp -\beta Ef \tag{1.89}$$

$$\frac{k_A}{k_{-A}} = \exp(-Ef) \tag{1.90}$$

can be inserted into (1.88) to produce

$$i = nFA \frac{k_A'}{k'_{-A}} C_A{}^b (C_x{}^b)^2 (1 - \theta_T) \exp[-(1 + \beta)fE]. \tag{1.91}$$

At low coverage the current should be second order in $C_x{}^b$, first order in $C_A{}^b$ and have a Tafel slope of 40 mV. If two steps are in equilibrium before the rate determining step the current becomes

$$i = nFA \frac{k_A'}{k'_{-A}} C_A{}^b (C_x{}^b)^3 (1 - \theta_T) \exp[-(2 + \beta)\alpha E] \tag{1.92}$$

that is the current should be third order in $C_x{}^b$, first order in $C_A{}^b$ and have a characteristic slope of 24 mV. The Langmuir isotherm has been used in this section to describe the relation between θ_A and $C_A{}^s$ However other isotherms are possible.

Temkin isotherm

Equilibrium for the first two reactions (1.81) and (1.82) written as

$$A + X \underset{k_b'}{\overset{k_f'}{\rightleftharpoons}} B_{ads} + e \tag{1.93}$$

gives

$$k_f' C_x (1 - \theta_T) \exp(\beta g\theta) = k_b' \theta_b \exp[-(1 - \beta)g\theta] \tag{1.94}$$

where g is a constant. The specific forms of the Temkin and Langmuir isotherms enter at equations (1.94) and (1.84). The former includes a term for interaction between adsorbed species[8,22]. It is usually assumed that $(1 - \theta_T)$ and θ_B are constant hence

$$C_x K' \exp(-Ef) = \exp(-g\theta) \tag{1.95}$$

where K' is another constant. Again if the rate determining step is

$$i = nFA k_f C_x{}^b \tag{1.96}$$

then

$$i = nFAk_1 \exp(-\beta fE) C_x{}^b K'' \tag{1.97}$$

therefore the current should be first order in C_x^b and have a Tafel slope of 120 mV. This is quite different from that expected for a Langmuir isotherm.

A number of reaction schemes have been characterised in this way but the exercise to date has been largely academic.

MORE COMPLICATED SURFACE REACTIONS

More complicated types of reaction have been considered in the literature. A well known one is the hydrogen evolution reaction. In principle once H has been formed

$$H^+ + e \rightleftharpoons H_{ads} \tag{1.98}$$

two alternative routes are open

$$H_{ads} + H^+ + e \rightleftharpoons H_2 \tag{1.99}$$

$$H_{ads} + H_{ads} \rightleftharpoons H_2. \tag{1.100}$$

The basic equations can be written down in a straightforward manner[23]. A detailed description of the results is given in the extensive literature.[8,22,31]

EXPERIMENTAL PROCEDURE

At this point the basic principle of some electrochemical methods will be described.

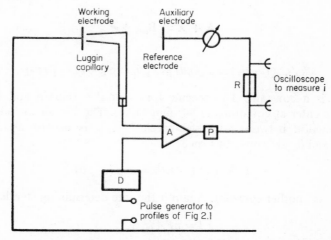

FIG. 1.6. A potentiostatic circuit. A is a wide band amplifier, amplification about 30,000, P is the power stage to deliver perhaps ± 3 A at ± 70 V. D is a potentiometer to provide a d.c. potential. R is a current measuring resistor.

POTENTIOSTATIC METHOD

The design of fast potentiostats and pulse gererators[24,25] is now well understood. Many types of instrument are commercially available. A versatile system can be built from operational amplifiers.[26,27,28]

A typical potentiotatic circuit is shown diagrammatically in Fig. 1.6, although there are other configurations. The current can be measured on a meter if a stationary measurement is intended, of oscillographically for a perturbation method. The current, as a signal across R must be measured with a differential input oscilloscope.

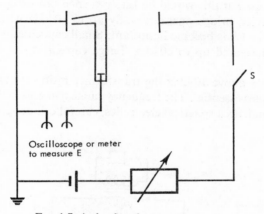

FIG. 1.7. A simple galvanostatic circuit.

GALVANOSTATIC METHOD

A galvanostatic circuit is shown in Fig 1.7. The resistance is made large enough so that a constant current flows irrespective of the cell characteristics. A mercury switch, S, starts the pulse. For fast response it is preferable to switch electronically, see Fig. 1.8.

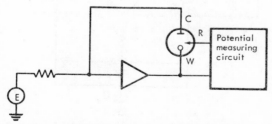

FIG. 1.8. Galvanostatic circuit powered by an operational amplifier.

A.C. METHOD

The Wien bridge shown in Fig. 1.9 has been widely used to measure the impedance as a series, R_s and C_s, or parallel, R_p and C_p, combination. A Wagner earthing system is also shown in the Fig. 1.9 to compensate for stray capacitance to earth. The need for a Wagner earth depends on the impedance being measured, but probably becomes important when[29] (a) measuring capacities <0.1 (b) for frequencies >5 kHz (c) measuring resistances at very high frequencies. Normally detection in order to adjust the signal at AB to zero at balance is by tuned amplifier (to eliminate harmonics) or phase sensitive detection. The signal appearing at AB can be directly measured on an ocilloscope but this would be rarely carried out except to get a crude balance. The Wien bridge operates easily up to 10 kHz when the effects of inductance in the leads become important. Small capacities where $(1/wC_s) > L_{\text{stray}}$ can be measured up to 60 kHz. Large capacities can become inaccurate at 1 kHz.

At frequencies above 10 kHz the transformer ratio arm bridge, shown in Fig. 1.10 becomes essential. The frequency range can reasonably be extended to 400 kHz. Small area spherical electrodes, careful cell design and allowance

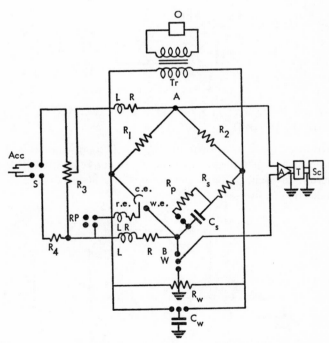

FIG. 1.9. A typical Wien bridge circuit with Wagner earth.[29]

for transmission line effects of the leads connecting the bridge to the cell are necessary.

Another type of measurement which has been used for a number of years, depends on measuring Lissajous figures.

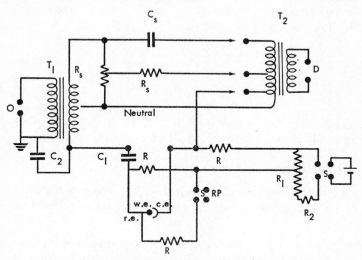

FIG. 1.10. Ratio arm bridge made by Wayne–Kerr B 601.

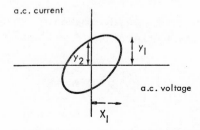

FIG. 1.11. Lissajous figure generated when an a.c. potential input to the cell is displayed against the output.

If an a.c. signal of small amplitude is fed into the pulse generator terminals of Fig. 1.6 then the output signal of Fig. 1.11 can be generated. In order to do this the input a.c. signal goes directly to the X plates of an oscilloscope and the output signal observed across R goes to the Y plates. The capacity and resistance R_s, C_s of the cell are related to the dimensions of the ellipse by

$$Z = \frac{x_1}{y_1} \tag{1.101}$$

$$\sin \phi = \frac{y_2}{y_1} \tag{1.102}$$

a is the amplitude of the a.c. signal. $\eta = a \sin (\omega t)$ and R is the measuring resistor in Fig. 1.16. If the impedance is split into the components R_s, C_s of the circuit shown in Fig. 1.11, then

$$R_s = (Z \cos \phi - R_c)A \tag{1.103}$$

$$\frac{1}{\omega C_s} = (Z \sin \phi)A. \tag{1.104}$$

The frequency range in this case is limited by the characteristic of the potentiostat. Typically R_s, C_s can be measured up to about 1–5 kHz.

REVIEW OF THE EXPERIMENTAL METHOD

(1) Obtain a current–voltage curve as in Fig. 1.1. The rotating disc, polarography or linear potential sweep at a stationary electrode would be recommended.

(2) Run a coulometry experiment at a suitable potential from (1) and analyse chemically for the products and reactant, check that reactant depletion

TABLE 1.1. Some examples which analyse stationary current–voltage curves (see Tanaka and Tamamushi[30], and Vetter[31]).

System	Method	Reference
$Zn(Hg)/Zn^{2+}$	Polarography	33
Fe^{2+}/Fe^{3+} on Pt	Rotating disc	34, 35
$Fe(CN)_6^{3-}/Fe(CN)_6^{4-}$ on Pt	Rotating disc	34
C^{3+}/C^{4+} on Pt	Rotating disc	36
V^{3+}/V^{2+} on Hg	Polarography	15
Cu^{2+}/Cu	Rotating disc	37
N_2H_4 on Pt	Rotating disc	38, 39
Cl_2/Cl^- on Pt	Rotating disc	40
Mn^{3+}/Mn^{2+} on Pt	Rotating disc	41
Mn^{4+}/Mn^{3+} on Pt	Rotating disc	41

matches product increase. There is a possibility that electrokinetic data can be obtained at the same time (see Chapter 2). The importance of chemical analysis and material balance cannot be over-emphasised, particularly since much published work is invalidated by its absence.

(3) Test the diffusion current observed in the measurements described paragraph (1). (see Fig. 1.1).

(4) This should match up with the concentration of the reactant in solution and have the correct characteristics depending on the method used for the measurement. (see Chapter 2). Extrapolate out diffusion at each potential by the rotating disc of potential step method as described in Chapter 2.

(5) Plot a current voltage curve without diffusion and examine by equations given in this chapter (1). Measure the order of reaction with rdspect to reactant, product, or intermediate, either by examining the effect on i at fixed E or by the effect on i_0.

(6) Examine by the non-steady state techniques of Chapter 2, in the order (a) fast linear sweep, (b) potentiostatic pulse. If the reaction is then well known accurate data can be obtained with the a.c. method

The above recipe is not infallible and depends very much on the system. Chemical intuition and common sence will largely dictate the sequence of events.

REFERENCES

1. Grahame, D. C. *Chem. Rev.* **41**, 441 (1947).
2. Parsons, R. "Modern Aspects of Electrochemistry", No. 1 (Ed., J. O. M. Bockris), Butterworths, London (1952).
3. Barlow, Jr, C. A., Macdonald, R. and Ross, J. *In* "Advances in Electrochemistry and Electrochemical Engineering", Vol. 6 (Ed., P. Delahay), Interscience, New York (1966).
4. Payne, R. *In* "Advances in Electrochemistry and Electrochemical Engineering", Vol. 7 (Ed., P. Delahay), Interscience, New York (1970).
5. "Physical Chemistry" (Eds., H. Eyring, D. Henderson and W. Jost), Vols. 9A, 9B. Academic Press, London and New York (1970).
6. Frumkin, A. N. and Damaskin, B. B. *In* "Modern Aspects of Electrochemistry" (Ed., J. O' M. Bockris), Butterworths, London (1964).
7(a). Frumkin, A. N. *Disc. Far. Soc.* **1**, 57 (1947).
7(b). Parsons, R. *Surf. Sci.* **2**, 418 (1964).
8. Delahay, P. "Double Layer and Electrode Kinetics", Interscience, New York (1965).
9. Sluyters, J. H. *Rec. Trav. Chim.* **79**, 1092 (1960).
10. de Levie, R. *In* "Advances in Electrochemistry and Electrochemical Engineering", Vol. 6 (Eds., P. Delahay and C. W. Tobias), Interscience, New York (1967).
11. Giles, R. D. and Harrison, J. A. *J. electroanal. Chem.* **24**, 399 (1970).

12. Parsons, R. *Trans. Far. Soc.* **47,** 1332 (1951).
13. Mohilner, D. M. and Delahay, P. *J. phys. Chem.* **67,** 588 (1963).
14. Marcus, R. A. *J. chem. Phys.* **43,** 679 (1965).
15. Randles, J. E. B. *Can. J. Chem.* **37,** 238 (1959).
16. Mairanovski, S. G. "Catalytic and Kinetic Waves in Polarography", Plenum Press, New York (1968).
17. Koryta, J. *In* "Advances in Electrochemistry and Electrochemical Engineering", Vol. 6 (Ed., P. Delahay), Interscience, New York (1967).
18(a). Parsons, R. *Surf. Sci.* **18,** 28 (1959).
18(b). Timmer, B., Sluyters-Rehbach, M. and Sluyters, J. H. *Surf. Sci.* **18,** 44 (1959).
19. Hale, J. M. *J. electroanal. Chem.* **8,** 181 (1964).
20. Hurd, R. M. *J. electrochem. Soc.* **109,** 327 (1962).
21. Piersma, B. J. and Gileadi, E. *In* "Modern Aspects of Electrochemistry", Vol 4. (Ed., J. O. M. Bockris), Plenum Press, New York (1966).
22. Conway, S. E. "Electrode Processes", Ronald, New York (1965).
23. Parsons, R. *Trans. Far. Soc.* **54,** 1053 (1958).
24. Brown, E. R., Smith, D. E. and Booman, G. L. *Anal. Chem.* **40,** 1411 (1968).
25. Brown, O. R. *Electrochim. Acta* **13,** 317 (1968).
26. Operational Amplifiers Symposium, *Anal. Chem.* **35,** 1770 (1963).
27. Goolsby, A. D. and Sawyer, D. T. *Anal. Chem.* **39,** 411 (1967).
28. Brand, M. J. D. and Fleet, B. *Chem. in Britain* 557 (1970).
29. Armstrong, R. D., Race, W. P. and Thirsk, H. R. *Electrochim. Acta* **13,** 215 (1968).
30. Tanaka, W. and Tamamushi, R. *Electrochim. Acta* **9,** 963 (1964).
31. Vetter, K. J. "Electrochemical Kinetics", Academic Press, London and New York (1967).
32. Gerischer, H. and Vielstich, W. *Z. phys. Chem.* **3,** 16 (1955).
33. Koryta, J. *Electrochim. Acta* **6,** 67 (1962).
34. Galus, Z. and Adams, R. N. *J. phys. Chem.* **67,** 866 (1963).
35. Angell, D. and Dickinson, T. (in press).
36. Vetter, K. J. *Z. phys. Chem.* **196,** 360 (1951).
37. Brown, O. R. and Thirsk, H. R. *Electrochim. Acta* **10,** 383 (1965).
38. Harrison, J. A. and Khan, Z. A. *J. electroanal. Chem.* **28,** 131 (1970).
39. Harrison, J. A. and Khan, Z. A. *J. electroanal. Chem.* **26,** 1 (1970).
40. Dickinson, T., Greef, R. and Wynne-Jones, L. *Electrochim. Acta* **14,** 467 (1969).
41. Vetter, K. J. and Manecke, G. *Z. phys. Chem.* **195,** 270 (1950).

Chapter 2

Non-stationary Methods

In Chapter 1, the factors which determine the observed current were discussed in general. It was also shown that the stationary voltage curve could be interpreted in terms of the kinetics in a number of cases. However in a current–voltage curve most of the information about the interfacial reaction is contained at the foot of the wave where measurement is inexact. In addition the maximum interfacial rate which can be measured in a stationary current–voltage curve (see diffusion layer model) depends on the diffusion layer thickness. Faster rates can be investigated by non-stationary methods in which the interfacial rate, which moves in sympathy with the signal, wins over processes occuring in solution. This amounts to setting up a thinner effective diffusion layer.

In this chapter the influence of diffusion on the measured parameter's current or potential, when either of these is perturbed, will be discussed. For many years the measurement of the characteristic parameters, i_0, α, for simple electron exchange reactions after correction for diffusion, was a major aim of electrochemistry. However, interest has turned to investigations of coupled electrochemical and chemical reactions, and surface reactions. This has been occasioned by renewed interest in electrosynthesis of organic compounds,[1,2] and by developments in fuel cell technology.[3,4,5] In Chapter 3 the kinetics of phase formation are described.

It must be said at the outset that not much space will be given to polarography. The reason for this is that although polarography has been a major field of activity, and is still important in chemical analysis, interest has switched to more versatile methods. There are a large number of text books available on this subject.[6,7,37,38] The theoretical results for polarography, however, will be similar to those given in this chapter, which will usually be for a planar, or spherical fixed area electrode. Polarography is less favoured at the moment for electrochemical kinetic investigations because:

(a) it is limited to low concentrations of reacting species, an obvious disadvantage for practical synthetic investigations;

(b) it is really limited to mercury;

(c) the interpretation of results is complicated by convection in the mercury, the polarographic maxima;

(d) it is limited to low values of i_0, and low rates of coupled chemical reactions;

(e) the changing surface area of the drop makes it difficult to combine with more appropriate potential–time profiles.

It must be said that an automatic polarograph is a very convenient tool for preliminary scanning experiments. Analysis of current–potential curves in suitable systems can also be made quite accurately if precautions are taken.[8] The continuous renewal of the mercury surface is an advantage; however, with the widespread use of pre-electrolysis and activated charcoal to clean solutions, similar accuracy can be obtained if necessary at stationary electrodes of other metals.

INTRODUCTION TO THE THEORY AND REACTION SCHEMES

If the electrochemical reaction proceeds at a reasonable rate the concentration of reacting species will drop. The rate of electrochemical reaction will be controlled both by the electrochemical rate and by the rate at which reactant arrives at the interface, that is by equation (1.14), which is written below as (2.1). Possible ways in which reactant is transported to the surface are by diffusion, convection, migration. Migration is invariably prevented by the addition of excess indifferent electrolyte. Natural convection can be avoided if the time of measurement is short. Both these effects will not be considered further.

The most important ways of coping with diffusion are perturbation methods. The reason for this is that the interfacial electrochemical reaction responds instantaneously to a change in potential, while the concentration of the reactant in solution changes slowly. For this reason extrapolation procedures can be used to remove the diffusion contribution to the electrochemical rate. The starting point for the calculation is equation (2.1), which is valid when both components of a redox couple are present in solution.

$$\frac{i}{A} = i_0 \left[\frac{C_O^s}{C_O^b} \exp\left(-\alpha f \eta\right) - \frac{C_R^s}{C_R^b} \exp\left[(1-\alpha)f\eta\right] \right] \qquad (2.1)$$

equally, this could be written for $(E - E_0)$ or E as the potential determining parameter, see Chapter 1. In principle, for any perturbation, C_O^s, C_R^s, are calculated from Fick's laws of diffusion and the appropriate boundary and initial conditions. Unfortunately there is no independent way of measuring

$C_O{}^s$, $C_R{}^s$, and they must be calculated theoretically from experimentally measureable quantities. It may be possible in the future to measure $C_O{}^s$, $C_R{}^s$ by a spectroscopic method, see Chapter 4. However as will become apparent in this Chapter a unified way of dealing with this problem emerges (see Table 2.3). Equation (2.1) then gives the operational dependence of current if the potential is perturbed, or potential if current is disturbed. Some common perturbation profiles are shown in Fig. 2.1. Each of these has its own distinct advantage which we will try to illustrate in this chapter. In the following text, reaction schemes will be discussed under headings denoting the perturbation method. The particular perturbation determines the boundary conditions for the solution of Fick's diffusion equations and in essence determines the form of the mathematical solution. The main intent is to present clearly the theoretical result and show how it is, or can be, used. The following reaction schemes or variations of them, can be handled at present:

(i) $O + ne \rightleftharpoons R$ electron transfer

(ii) $Y \rightleftharpoons O$ preceding chemical reaction
$O + ne \rightleftharpoons R$

(iii) $O + ne \rightleftharpoons R$ catalytic reaction
$R \rightarrow O + B$

(iv) $O + ne \rightleftharpoons R$ E.C. reaction
$R \rightarrow B$

(v) $O + ne \rightleftharpoons R$ E.C.E. reaction
$R \rightarrow P$
$P + n_2 e \rightleftharpoons B$

(vi) $O + ne \rightleftharpoons R$ dimerisation
$R + R \rightarrow R_2$

(vii) $O + ne \rightleftharpoons R$ polymerisation
$R + O \rightarrow RO$
$RO + O \rightarrow RO_2$
$RO_n \rightarrow B$

(viii) $O \rightleftharpoons O_{ads}$ parallel surface and solution redox reaction
$O_{ads} + ne \rightarrow B_1$
$O + ne \rightarrow B_2$

(ix) $O + ne \rightleftharpoons R$ rate determining adsorption–desorption
$R \rightleftharpoons R_{ads}$

(x) $O \rightleftharpoons O_{ads}$ adsorption

(xi) $Y \rightleftharpoons O$ preceding reaction and adsorption
$O \rightleftharpoons O_{ads}$

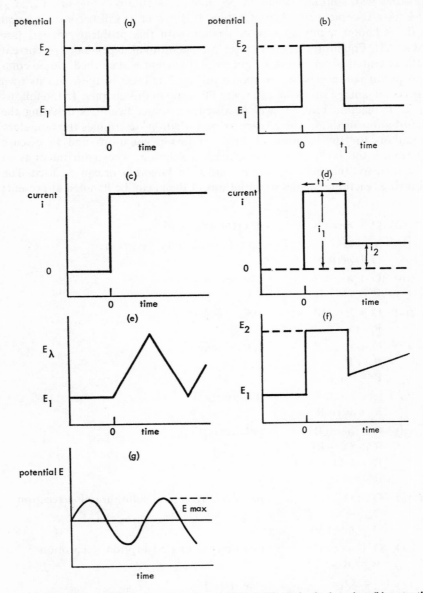

FIG. 2.1. Some perturbations in common use: (a) potentiostatic single pulse, (b) potentiostatic double pulse, (c) galvanostatic single pulse, (d) galvanostatic double pulse, (e) linear potential sweep, (f) potentiostatic pulse and sweep combined, (g) sine wave.

Some of these at the potentiostatic pulse or rotating disc have a simple mathematical solution, however the usage of digital computers means that any reaction scheme will be capable of solution. The sections devoted to different elcetrochemical methods are labelled (a) to (l). In all sections the numbers (i) to (viii) correspond to the reaction schemes listed above.

The complete diffusion equation with boundary and initial conditions will be written for different reaction schemes in Section (a) only. They can easily be written for any of the Sections (b) to (l) in a similar manner.

Setting up diffusion equations

Species which diffuse freely in solution obey Fick's first and second laws which are respectively

$$\text{flux} = D\left(\frac{\partial C}{\partial x}\right) \qquad \text{Fick's first law} \qquad (2.2)$$

$$\frac{\partial D}{\partial t} = D\left(\frac{\partial^2 C}{\partial x^2}\right) \qquad \text{Fick's second law.} \qquad (2.3)$$

Consider the response of the simple irreversible system

$$O + ne \overset{k_f}{\to} R$$

to a constant potential from time $t = 0$. The problem, usually, is to solve for C, the concentration, from Fick's second law with two boundary conditions

$$x = 0, \qquad \frac{i}{nA} = D\left(\frac{\partial C}{\partial x}\right)_{x=0} = k_f(C)_{x=0} \qquad (2.4)$$

$$x \to \infty, \qquad C \to C^b \qquad (2.5)$$

k_f is a constant given by perhaps

$$k_f = k_1 \exp\left(-\alpha nfE\right)$$

(see Chapter 1, equations (1.4), (1.9), or (1.12)). The experiment then corresponds to the potentiostatic pulse, profile (a) of Fig. 2.1, applied at $t = 0$. This system is described further in Section a(i). The state of the system before electrolysis, the initial condition, must also be given

$$t = 0, \quad C = C^b \quad \text{for all } x. \qquad (2.6)$$

Laplace transform method of solution

The Laplace transform of C with respect to t defined by

$$\bar{C} = \int_0^\infty \exp\left(-st\right) C \, dt \tag{2.7}$$

(see any standard text on the properties of the transform[9,10]) of both terms in equation (2.3) converts it from a partial differential equation to an ordinary differential equation with constant coefficients

$$\bar{C} - s(C)_{t=0} = D\left(\frac{\partial^2 \bar{C}}{\partial x^2}\right) \tag{2.8}$$

therefore

$$\bar{C} - sC^b = D\frac{\partial^2 \bar{C}}{\partial x^2} \tag{2.9}$$

An important feature of this equation is that the initial condition, equation (2.6), has been absorbed. The solution is

$$\bar{C} = A' \exp\left[-\left(\frac{s}{D}\right)^{1/2} x\right] + B' \exp\left[\left(\frac{s}{D}\right)^{1/2} x\right] + \frac{C^b}{s}. \tag{2.10}$$

The boundary condition at $x \to \infty$ leads immediately to $B' = 0$. The solution now becomes

$$\bar{C} = A' \exp\left[-\left(\frac{s}{D}\right)^{1/2} x\right] + \frac{C^b}{s} \tag{2.11}$$

hence

$$(\bar{C})_{x=0} = A' + \frac{C^b}{s}. \tag{2.12}$$

Only the conditions at $x = 0$ remains. The equation (2.11) is perfectly general, depending only on Fick's second law, the concentration at $t = 0$ and the concentration at $x \to \infty$.

Potentiostatic pulse

At this point the electrochemical technique must be introduced into the boundary condition at $x = 0$. For the potentiostatic pulse of height E starting at $t = 0$ (profile (a) in Fig. 2.1). k_f in equation (2.4) is a constant for times greater than zero, therefore

$$\frac{\bar{i}}{nFA} = D\left(\frac{\partial \bar{C}}{\partial x}\right)_{x=0} = k_f\left(A' + \frac{C^b}{s}\right) \tag{2.13}$$

$\bar{i}$ is the Laplace transform of the current.

Differentiating equation (2.11) gives

$$\frac{i}{nFA} = -A'(SD)^{1/2} = k_f\left(A' + \frac{C^b}{s}\right). \tag{2.14}$$

Rearrangement of the right hand terms leads to

$$A' = -\frac{k_f C^b}{s[(sD)^{1/2} + k_f]} \tag{2.15}$$

hence

$$\frac{i}{A} = \frac{nFk_f C^b}{s^{1/2}[(sD)^{1/2} + k_f]} \tag{2.16}$$

Transforming back to a function of time by referring to tables of Laplace transforms

$$\frac{i}{A} = nFk_f C^b \exp\left[\left(\frac{k_f}{D}\right)^2 t\right] \operatorname{erfc}\left[\left(\frac{k_f}{D}\right)t^{1/2}\right] \tag{2.17}$$

When the electrochemical reaction is reversible and k_f, k_b both need to be considered, a similar expression to equation (2.17) is obtained but with different constants in the exponents, see equation (2.33). The simplest case when only species O diffuses has been considered here for clarity.

Other techniques which require different boundary conditions at $x = 0$ can all be calculated by similar procedure. In some cases special mathematical techniques are needed, but almost invariably the starting point is equation (2.11) and (2.12).

A further point needs to be made about the Laplace method. Although it can be used, as here, as a cook book method, it is essential to check that the final equation obeys the initial differential equation. In this case the equation for the concentration distribution could be differentiated to ascertain that it obeys Fick's second law.

Potentiostatic sweep

A great deal of theoretical activity has been expended on this technique because of its usefulness. Assume once again that only O diffuses and the electrochemical reaction is irreversible. The extension to O, R diffusing and a reversible electrochemical reaction is easily accomplished. The equations to be solved are (2.4), (2.5), (2.6). Consider equation (2.12). The value of A'

can be found by differentiating equation (2.11)

$$\frac{i}{nFA} = \phi(s) = D\left(\frac{\partial \overline{C}}{\partial x}\right)_{x=0} = A'\sqrt{D_0 s} \qquad (2.18)$$

therefore

$$A' = -\frac{\phi(s)}{\sqrt{D_0 s}} . \qquad (2.19)$$

Equation (2.12) can now be inverted

$$C_O{}^s = (C_O)_{x=0} = C_O{}^b - \frac{1}{\sqrt{D_0 \pi}} \int_0^t \frac{\phi(Z)dZ}{\sqrt{(t-Z)}} \qquad (2.20)$$

The expression for the current is

$$\frac{i}{nF} = \phi(t) = Ak_{sh} C_O{}^s \exp\left[-(E-E_0)\alpha nf\right]. \qquad (2.21)$$

Suppose in a sweep experiment

$$E = E_i - vt \qquad (2.22)$$

$$b = \alpha nfv \qquad (2.23)$$

where E_i is a starting potential and v the sweep rate in volts $\sec^{-1}$

$$\phi(t) = Ak_{sh}\left[C_O{}^b - \frac{1}{\sqrt{D_0 \pi}} \int_0^t \frac{\phi(Z)dZ}{\sqrt{(t-Z)}}\right] \exp\left[-(E_i-E_0)\alpha nf\right] \exp(bt). \qquad (2.24)$$

In order to make the parameters non-dimensional the following changes are made

$$\chi(bt) = \frac{\phi(t)}{AC_O{}^b\sqrt{\pi D_O b}} = \frac{i}{nFAC_O{}^b\sqrt{\pi D_O b}} \qquad (2.25)$$

$$\exp u = \frac{\sqrt{\pi D_O{}^b}}{k_{sh}} \exp\left[(E_i - E_0)\alpha nf\right]. \qquad (2.26)$$

The final equation is

$$1 - \int_0^{bt} \frac{\chi(Z)dZ}{\sqrt{bt - Z}} = \exp(u - bt)\chi(bt). \qquad (2.27)$$

Equation (2.27) is a Volterra integral equation of the second kind and can be calculated numerically.[47] If u is given, $\chi(bt)$ hence i can be calculated as a function of bt. A discussion of the numerical methods available for solving equations of the type (2.27) which appear in linear sweep has been given in the literature.[11]

As the flux of concentration $D(\partial C/\partial x)$ at $x = 0$ is in general required, it is in a sense superfluous to calculate the concentration distribution given by equation (2.13). Recently a mathematical procedure has been devised which makes it possible to calculate $(\partial C/\partial x)_{x=0}$ directly.[12] However this method has not yet been developed and its range of application is not yet certain.

Non-dimensional form of the diffusion equations

It will be noticed in the sweep section that some nondimensional parameters were introduced. This is a general principle when dealing with numerical methods. The time of calculation can be drastically reduced and comparison of experiment and theory becomes much simpler. Many examples of how experiment and theory can be compared will be given in the rest of the chapter. In this section a straightforward diffusion equation will be transformed into non-dimensional form. As a concrete example consider the catalytic reaction described further on p. 40. The differential equation for spherical diffusion to a stationary electrode, describing the response of the concentration of the species A in the scheme

$$2A + 2e \rightleftharpoons 2B$$

$$2B \xrightarrow{k_2} A + C \tag{2.28}$$

is

$$\frac{\partial C_A}{\partial t} = D_A \left(\frac{\partial^2 C_A}{\partial r^2} + \frac{2}{r} \frac{\partial C_A}{\partial r} \right) + \tfrac{1}{2} k_2 C_B^2. \tag{2.29}$$

The first term in the brackets in equation (2.29) is Fick's second law, equation (2.3), written for spherical diffusion.

Applying the transformations[13]

$$\psi = \frac{r}{r_0} \left(1 - \frac{2C_A + C_B}{2C_A^b} \right) \tag{2.30}$$

$$R = \frac{r - r_0}{D^{1/2} t^{1/2}} \tag{2.31}$$

$$T = 2k_2 C_A^b t \tag{2.32}$$

gives the non-dimensional form of equation (2.29) as

$$\frac{\partial \psi}{\partial T} = \frac{1}{T} \frac{\partial^2 \psi}{\partial R^2} + \frac{R}{2T} \frac{\partial \psi}{\partial R} \qquad (2.33)$$

and this could be the starting point for a numerical method of solution, discussed in the next section. Transformations like (2.30), (2.31), (2.32) are very common.

Numerical method of solution

The reader is referred to texts on numerical analysis for a detailed account. In general the particular problem dictates the mathematical method. However the finite difference method has been well tried in electrochemistry and will be discribed here. It is doubtful if the method can in practice be as universally applied as suggested in the literature. A comprehensive review of the method is available.[14]

The form of Fick's second law of linear diffusion

$$\frac{\partial C}{\partial t} = D \frac{\partial^2 C}{\partial x^2}$$

can be written by reference to Fig. 2.2 as

$$\frac{\bar{C}_n - C_n}{\Delta t} = \frac{D}{(\Delta_x)^2} [C_{n+1} - 2C_n + C_{n-1}]. \qquad (2.34)$$

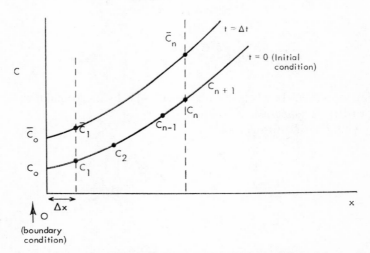

FIG. 2.2. Principle of the finite difference method of solving differential equations.

It can be shown for this particular formulation the solution is stable if

$$\frac{\Delta t D}{(\Delta x)^2} \leqslant \frac{1}{6}.$$ (2.35)

Introduction of this value into equation (2.34) gives

$$\bar{C}_n = \frac{1}{6}[C_{n+1} + 4C_n + C_{n-1}].$$ (2.36)

The starting point to the calculation is the initial condition which defines a set C_n at $t = 0$. It these are used three at a time, as equation (2.36), a new set C_n at $t = \Delta t$ can be calculated. The points $\bar{C}_0$ and $\bar{C}_n$ are known from the boundary conditions. In this way approximate concentration distance profiles can be calculated for $t = \Delta t, 2\Delta t, 3\Delta t$, etc. The corresponding current at each time is given by

$$i = DnFA \frac{(C_1 - C_0)}{\Delta x}.$$ (2.37)

It has been suggested that this constitutes a general method for solving the differential equations met in electrochemistry. In a similar way to equation (2.34), equations for linear sweep, rotating disc, polarography potentiostatic pules can easily be written for almost any reaction scheme. There are other types of the finite difference method with appropriate mathematical stability requirements.

TECHNIQUES AND REACTION SCHEMES

In the first section, concerned with the potential step method, the equations to be solved will be written out in full.

POTENTIAL STEP METHOD

(i) *Electron transfer*

Consider the straight electron exchange reaction[15] in which both species are diffusing to or away from the electrode

$$O + ne \rightleftharpoons R$$

The equations to be solved are

$$\frac{\partial C_O}{\partial t} = D_O \left(\frac{\partial^2 C_O}{\partial x^2} \right)$$ (2.38)

$$\frac{\partial C_R}{\partial t} = D_R \left(\frac{\partial^2 C_R}{\partial x^2} \right)$$ (2.39)

with the boundary conditions at $x = 0$

$$i = nFA \left(\frac{\partial C_O}{\partial x}\right) = nFA \left(k_f C_O^s - k_b C_R^s\right) \quad (2.40)$$

$$D_O \left(\frac{\partial C_O}{\partial x}\right)_{x=0} = - D_R \left(\frac{\partial C_R}{\partial x}\right)_{x=0} \quad (2.41)$$

with the conditions at $x \to \infty$

$$C_O = C_O^b, \qquad C_R = C_R^b \quad (2.42)$$

and the initial condition at $t = 0$

$$C_O = C_O^b, \qquad C_R = C_R^b \quad (2.43)$$

The well known solution is a falling transient with time, shown in Fig. 2.3, given by

$$\frac{i}{A} = i_0 \left\{\exp\left(- \alpha f\eta\right) - \exp\left[(1 - \alpha)f\eta\right]\right\} \exp \lambda^2 t \, \text{erfc} \, \lambda t^{1/2} \quad (2.44)$$

$$\frac{i}{A} = I \exp \lambda^2 t \, \text{erfc} \, \lambda t^{1/2} \quad (2.45)$$

where I is the current free of diffusion, and the term in brackets modifies it. λ is given by

$$\lambda = \frac{i_0}{nF} \left\{\frac{\exp\left(- \alpha f\eta\right)}{C_O^b D_O^{1/2}} + \frac{\exp\left[(1 - \alpha)f\eta\right]}{C_R^b D_R^{1/2}}\right\} \quad (2.46)$$

Equation (2.44), which appears to be rather complex. is rarely used in this

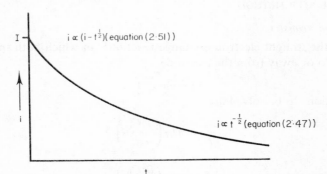

FIG. 2.3. Potentiostatic current time transient given by equation (2.44) (diagrammatic).

form. Two limiting cases can be calculated, by expanding the exp and erfc terms in equation (2.44). When $\lambda t^{1/2} > 5$

$$\frac{i}{A} = I/\pi^{1/2}\lambda t^{1/2}. \tag{2.47}$$

This equation, when I is given by

$$I = i_0 \{ \exp(-\alpha f\eta) - \exp[(1-\alpha)f\eta] \} \tag{2.48}$$

and λ from equation (2.46) are substituted, is independent of α, i_0 and is completely diffusion controlled. Either i_0 or t can be large to cause this situation. A commonly occurring situation corresponds to $C_R \to \infty$, for example when species R is a metal then equation (2.44) becomes

$$i = nFA\,C_O{}^b \left(\frac{D}{\pi t}\right)^{1/2} [1 - \exp(-nf\eta)]. \tag{2.49}$$

At high negative potential equations (2.45) and (2.49) become

$$i = nFA\,C_O{}^b \left(\frac{D}{\pi t}\right)^{1/2} \tag{2.50}$$

Equation (2.50) is very important and must always be obeyed in the limit. It is usual to test experimental data by plotting i versus $t^{-1/2}$ at negative potentials. If the slope corresponds to about the expected value from equation (2.50) then if n is known from coulometry an accurate D can be calculated. Catalytic (higher slope than diffusion) and slow preceding chemical reactions (lower slope than expected) can be detected by this method. With the opposite assumption, $\lambda t^{1/2} < 1$, equation (2.44) becomes

$$\frac{i}{A} = I\left(1 - \frac{2\lambda t^{1/2}}{\pi^{1/2}}\right) \tag{2.51}$$

which means that I can be measured by plotting i against $t^{1/2}$ and extrapolating to $t = 0$ as shown in Fig. 2.3. The cross over from equation (2.47) to equation (2.51) is discussed in the literature.[16] When the electrochemical reaction is very slow $\lambda t^{1/2} \ll 1$ then

$$\frac{i}{A} = I \tag{2.52}$$

and the transient is a constant current. If the experimental curve fits equations (2.51), (2.47) or (2.49) a Tafel slope, $\log I/E$, can be drawn.

A simple redox reaction should show the behaviour of Fig. 2.6 if a double pulse, profile, Fig. 2.1(b), is applied to the electrode.

(ii) First order preceding chemical reaction

$$Y \underset{k_2}{\overset{k_1}{\rightleftharpoons}} O$$

$$O + ne \rightleftharpoons R$$

can be dealt with in a similar way to equation (2.44). If it is assumed that O is immediately reduced on reaching the electrode then[17,18,41] the limiting current is given by

$$i = nFA(C_y{}^b + C_O{}^b)D^{1/2}k_1{}^{1/2}K^{1/2} \exp(k_1 Kt) \operatorname{erfc}(k_1 Kt)^{1/2}. \tag{2.53}$$

The form of the current–time curve is similar to Fig. 2.3. Y and O are present in the bulk at concentrations $C_Y{}^b$, $C_O{}^b$. This equation has two limits

$$t < \frac{1}{k_1 K}; \quad i = nFA(C_Y{}^b + C_O{}^b)(Dk_1 K)^{1/2} \left[1 - \frac{2}{\pi^{1/2}}(k_1 Kt)^{1/2}\right] \tag{2.54}$$

$$t > \frac{1}{k_1 K}; \quad i = nFA(C_Y + C_O{}^b)\left(\frac{D}{\pi t}\right). \tag{2.55}$$

For details of calculating k, see ref. 17. The pulse method has rarely been used to measure k_1.

The a.c. method and rotating disc have been much more widely used for this purpose.

(iii) Catalytic reaction

The catalytic reaction of second order

$$2A + 2e \rightleftharpoons 2B$$

$$2B \overset{k_2}{\rightarrow} A + C \tag{2.56}$$

and the first order variant

$$A + e \rightleftharpoons B$$

$$B \overset{k_1}{\rightarrow} A + C \tag{2.57}$$

where C is electroinactive, have been described by Booman and Pence.[13]

The problem for reaction scheme (2.56) is defined by the spherical diffusion equations

$$\frac{\partial C_A}{\partial t} = D_A \left(\frac{\partial^2 C_A}{\partial r^2} + \frac{2}{r} \frac{\partial C_A}{\partial r} \right) + \frac{1}{2} k_2 C_B^2 \tag{2.58}$$

$$\frac{\partial C_B}{\partial t} = D_B \left(\frac{\partial^2 C_B}{\partial r^2} + \frac{2}{r} \frac{\partial C_B}{\partial r} \right) - k_2 C_B^2 \tag{2.59}$$

with the conditions at $r = r_0$

$$C_A = 0 \tag{2.60}$$

$$D_A \left(\frac{\partial C_A}{\partial r} \right) = - D_B \left(\frac{\partial C_B}{\partial r} \right) \tag{2.61}$$

and with the condition at $r \to \infty$

$$C_A \to C_A^b, \qquad C_B \to 0. \tag{2.62}$$

The initial condition is

$$C_A = C_A^b, \qquad C_B = 0 \tag{2.63}$$

The original form of the equations has been retained and the equations are written for spherical diffusion. The second order scheme is solved numeri-

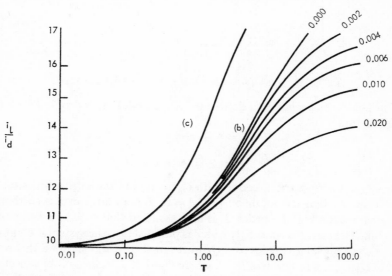

FIG. 2.4. Working curve i_l/i_d versus T for (a) first order, (b) second order catalytic reaction as a function of Q.[13]

cally and the first order analytically. The results are shown in Fig. 2.4. It is assumed that $D_A = D_B$, T, Q are dimensionless variables given by

$$T = k_1 t \tag{2.64}$$

$$Q = \frac{1}{r_0} \left(\frac{D}{2 C_A{}^b k} \right)^{1/2} \tag{2.65}$$

t is real time. Simple curve fitting of the experimental $i(t)$ transient to Fig. 2.4 should give D and k. i_d must be estimated or measured at short times where $i/i_d \to 1$.

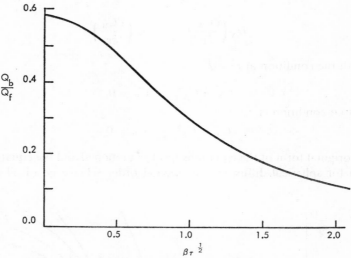

FIG. 2.5. Q_b/Q_f as a function of $\beta \tau^{1/2}$ for a catalytic reaction.[19]

Christie[19] has suggested a double pulse method, that is profile Fig. 2.1(b). The reaction

$$A + e \rightleftharpoons B$$

$$B + Z \xrightarrow{k_1} A + C \tag{2.66}$$

is driven in the forward direction with $C_A{}^s = 0$ until time τ then the situation is reversed so that instantaneously $C_B{}^s = 0$. Z is in large excess so that the chemical reaction is first order. It is also assumed that species B is absent in the bulk. The forward charge Q_f which is pumped into the system is compared to the charge which is recovered Q_b in reversal. Fig. 2.5 shows a theoretical plot of Q_f/Q_b as a function of $\beta \tau^{1/2}$, where $\beta = k_1 C_Z$. Values of k_1 can clearly be obtained by comparison of Fig. 2.5. and the experimental data. Fig. 2.5 is to be compared to Fig. 2.6 for a simple redox reaction.

There is a conceptual difficulty[20,21] because equation (2.56), for example, implies that

$$B + e \rightleftharpoons C \qquad (2.67)$$

proceeds more easily than

$$A + e \rightleftharpoons B \qquad (2.68)$$

the question then arises why does not

$$A + 2e \rightleftharpoons C \qquad (2.69)$$

One assumes implicitly in writing equation (2.56) that

$$B + e \rightleftharpoons C \qquad (2.70)$$

is kinetically hindered. A number of other possibilities have been investigated by Feldberg[20] who shows that they cannot be distinguished easily from equation (2.56).

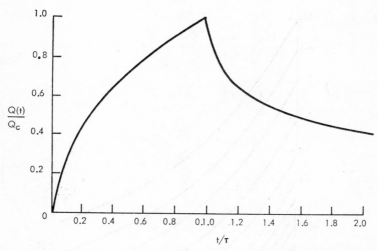

FIG. 2.6. Q_t/Q_c as a function of t/τ for a simple redox reaction.[19]

(iv) *E.C. reaction*

The E.C. mechanism

$$O + ne \rightleftharpoons R \qquad (2.71)$$

$$R \xrightarrow{k} B \qquad (2.72)$$

can be investigated[22] by a double potentiostatic step experiment. The current at a time t_1 during the forward reaction can be compared with a current at

$\tau + t_1$ during the back reaction. A theoretical calibration graph can be produced by solving the equations corresponding to scheme (2.72):

$$\frac{\partial C_O}{\partial t} = D_O \frac{\partial^2 C_O}{\partial x^2} \tag{2.73}$$

$$\frac{\partial C_R}{\partial t} = D_R \frac{\partial^2 C_R}{\partial x^2} - kC_R \tag{2.74}$$

with the boundary conditions at $x = 0$

$$0 < t < \tau \qquad C_O = 0 \tag{2.75}$$

$$t > \tau \qquad C_R = 0 \tag{2.76}$$

$$D_O \left(\frac{\partial C_O}{\partial x} \right) = -D_R \left(\frac{\partial C_R}{\partial x} \right) \tag{2.77}$$

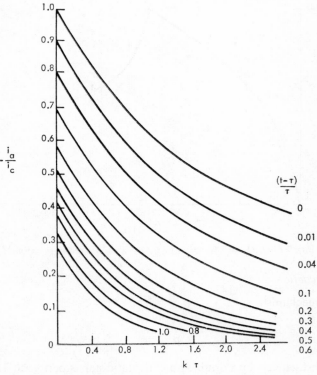

FIG. 2.7. i_a/i_c as a function of $k\tau$ for the E.C. mechanism.[22]

$$x \to \infty \quad C_O \to C_O{}^b$$
$$C_R \overset{k}{\to} 0 \tag{2.78}$$

and the initial condition

$$C_O = C_O{}^b; \quad C_R = O \tag{2.79}$$

Fig. 2.7 shows i_a/i_c as a function of $k\tau$ and $(t - \tau)/\tau$. The currents i_c and i_a are measured at equal times from the start of the electrolysis t_1 and from the switching point respectively.

A similar method[19] has been suggested in which the forward and reverse total charges are compared, see Fig. 2.8. The calculation uses the same initial and boundary conditions as ref. 22. However, the current comparison method is undoubtably more accurate.

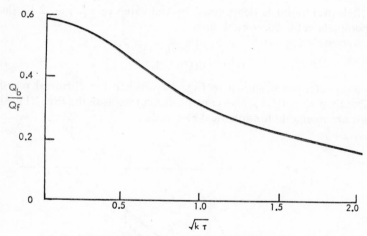

FIG. 2.8. Q_b/Q_f against $\sqrt{k\tau}$ for the E.C. mechanism.[19]

(v) *E.C.E. reaction*

The E.C.E. mechanism

$$O + n_2 e \rightleftharpoons R$$
$$R \underset{k'_b}{\overset{k'_f}{\rightleftharpoons}} P \tag{2.80}$$
$$P + n_2 e \rightleftharpoons B$$

can be investigated[23] by the following the current or charge with time. If the potentiostatic pulse is arrange so that the flux O and P are completely diffusion controlled, then for an irreversible chemical reaction the curves are only a function of k_f and D. The species R, P are absent in the bulk. The familiar

$$i = n_1 FA\, C_O^b \left(\frac{D}{\pi t}\right)^{1/2}$$

(2.81)

plot discussed on p. 39 for the reaction $O + ne \rightleftharpoons R$ without chemical complication now behaves as in Fig. 2.9. The limiting slope, at short times before R and the chemical reaction have built up, corresponds to n_1. At long times the slope switches over to

$$i = (n_1 + n_2)\, FAC_O^b \left(\frac{D}{\pi t}\right)^{1/2}.$$

(2.82)

The switch-over point is determined by the value of k_f. Fig. 2.10 shows a more complete calibration method.

An alternative approach[24] is to plot

$$(Q/t^{1/2})/(Q/t^{1/2})_{t \to \infty}$$

(2.83)

as the parameter, as is shown in Fig. 2.11. When the chemical reaction is reversible then $K = k_b'/k_f'$ also controls the curves as in the Fig. 2.11. Similar relations are available for spherical electrodes.

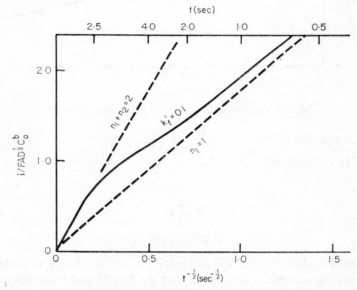

FIG. 2.9. The behaviour of an E.C.E. mechanism.[23]

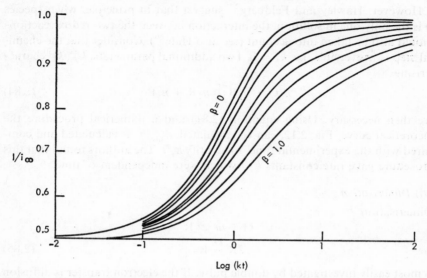

FIG. 2.10. General calibration curve for an E.C.E. mechanism at a spherical electrode.[23]

$$\rho = \frac{n_2}{n_1 + n_2} \qquad \beta = \frac{Dt}{r_0^2}$$

curves drawn for $\beta = 0, 0\cdot001, 0\cdot01, 0\cdot05, 0\cdot10, 0\cdot20, 0\cdot50, 1\cdot00$.

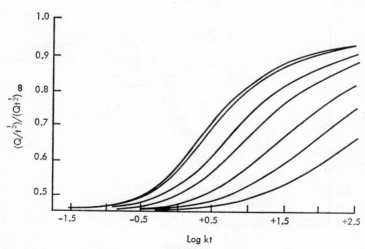

FIG. 2.11. Working curve for E.C.E. mechanism $n_1 = n_2$. From top to bottom; $K = 0\cdot1$, $0\cdot95, 2\cdot0, 5\cdot0, 10\cdot0, 20\cdot0$.[23]

However, Hawley and Feldberg[25] suggest that in principle, when species O is the only one in solution, the interaction between the two redox reactions should also be taken into account (see also Hale[26]). Consider that the chemical step is irreversible, i.e. $k_b = 0$. Two additional parameters, k_f', k_b' characterising

$$n_2O + n_1B \underset{k_b'}{\overset{k_f'}{\rightleftharpoons}} n_2R + n_1P \tag{2.84}$$

are then necessary. Using equation (2.81) and a numerical procedure the theoretical curve, Fig. 2.12, can be calculated. n_{app}/n_1 is calculated and compared with the experimental time, to identify $k_f't$. The authors found that this procedure gave rate constants $k_f't$ which were independent of time.

(vi) *Dimerisation*

Dimerisation

$$O + ne \rightleftharpoons R$$

$$2R \overset{k_2}{\rightarrow} R_2 \tag{2.85}$$

is most easily investigated by double pulse. If the electron transfer is diffusion controlled the electrical variable is removed and the current depends only on k_2, D. Species R is absent in the bulk of solution. In the original paper of Olmstead and Nicholson[27] the following non-dimensional quantities are defined

$$y = k_2C_O^b t$$
$$y_s = k_2C_O^b \tau \tag{2.86}$$

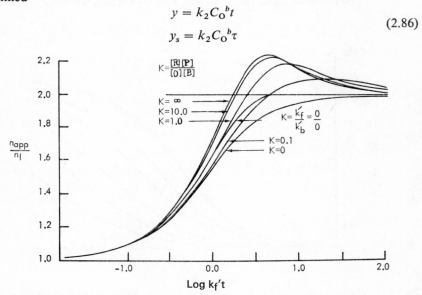

FIG. 2.12. n_{app}/n_1 as a function of kt for the E.C.E. mechanism.[25]

where τ is the switching time and y_s its non-dimensional equivalent. The current can then be defined by

$$i = nFA(k_2 C_O{}^b D_O)^{1/2} C_O{}^b \phi(y). \tag{2.87}$$

The authors calculate $\phi(y)$ numerically and give a table of $\phi(y)$ as a function of both y/y_s and y_s. With the help of this table experiment and theory can be compared. Fig. 2.13 shows theoretical non-dimensional current–time curves for first order chemical reaction and dimerisation compared to free diffusion of R.

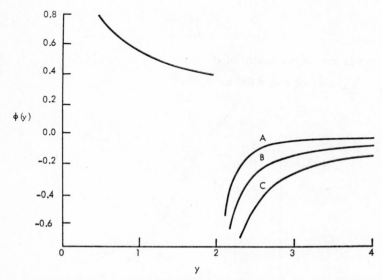

FIG. 2.13. Current as a function of non-dimensional time for (A) first order E.C., (B) dimerisation, (C) simple redox reaction.[27]

(viii) *Parallel surface and solution redox reaction*

A variation of the surface reaction given as equation (viii) in the introduction to this chapter has been investigated in the literature,[28] namely

$$A \rightleftharpoons A_{ads}$$
$$A_{ads} + e \rightleftharpoons B_{ads} \xrightarrow{k_{het}} C_{ads}$$
$$C_{ads} + Ze \rightarrow D_{ads} \tag{2.88}$$
$$B \rightleftharpoons B_{ads}$$
$$D \rightleftharpoons D_{ads}$$

B is absent in solution and initially the coverage with B is zero. The reaction is equivalent to the E.C.E. mechanism discussed in Section (v). If the potential is adjusted so that the step $A_{ads} + e \rightleftharpoons B_{ads}$ is completely diffusion controlled the system of equations can be solved easily by the Laplace method. Experiment and theory are most easily compared in a double pulse experiment by comparing charges in the forward $(Q_f = |-Q|,$ in the original work) and reverse direction $(Q_b = |+Q|)$. A calibration graph is shown in Fig. 2.14, where the parameter $\lambda_2^2 t_1$ is plotted against Q_b/Q_f. t_1 is the switching time. Moving the experiment data along the $\lambda_2^2 t_1$ axis allows λ_2 to be identified.

$$\lambda_2 = \frac{\gamma_2 k_{het}}{D_B^{1/2}} \tag{2.89}$$

γ_2 is the adsorption coefficient of B.

$C_B^s \gamma_2 = (C_B)$ surface and γ has units of cm.

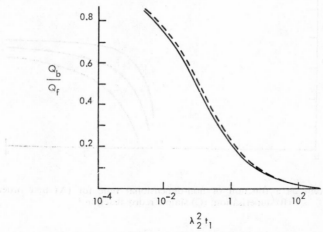

FIG. 2.14. Q_b/Q_f plotted as a function of $\lambda_2^2 t_1$ for a surface E.C.E. mechanism.[28]

CURRENT STEP METHOD; GALVANOSTATIC METHOD OR CHRONOPOTENTIOMETRY

The experimental equipment is relatively simple and has been discussed in Chapter 1.

The fact the potentiostats are easily available commercially means that interest in the galvanostatic method has waned. However the method has some advantages:

(1) The amount of charge flowing is easily measured as $I \times t$.

(2) The theoretical results are simple for a straightforward charge transfer.

(3) An overall picture is gained, in that in a single transient the whole potential range is covered. (This advantage has been lost to the potentiostatic linear sweep technique.)

The disadvantages usually outweigh the advantages, namely:

(4) The double layer charging interferes as it proceeds at the same time as the electrode reaction.

(5) The theory is difficult except for simple charge transfer.

The results for different reaction schemes will be very briefly reviewed as a large number of these are to be found in the older literature.

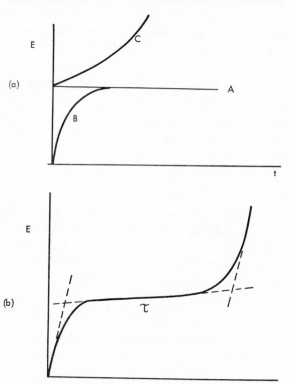

FIG. 2.15. Result of applying a constant current pulse to an electrode for a simple redox reaction (diagrammatic); τ is the transition time.
    (a) In the early stages:
        A. slow electrochemical step
        B. double layer charging
        C. with diffusion
    (b) Complete wave showing the transition time.

(i) *Electron transfer*

Simple charge transfer

$$O + ne \rightleftharpoons R$$

behaves as shown in Fig. 2.15. The transition time τ shown in the diagram is a characteristic quantity and in this case is given by the Sands equation which is derived directly from equation (2.13). The equation is valid if O, R or only O are present in the bulk.

$$i\tau_d^{1/2} = \frac{nFA(\pi D_O)^{1/2}C_O^b}{2}. \tag{2.90}$$

The equation holds for a reversible or irreversible charge transfer.

According to Gierst,[29] $i\tau^{1/2}$ plotted against i is a criterion that the system is a simple electron transfer consistent with equation (2.90).

A completely reversible redox reaction should in addition have the time dependence

$$E = E_0 + \frac{RT}{nF} \ln \left(\frac{D_R}{D_O} \right)^{1/2} + \frac{RT}{nF} \ln \left(\frac{\tau^{1/2} - t^{1/2}}{\tau^{1/2}} \right) \tag{2.91}$$

assuming that $C_R = 0$ at $t = 0$. An irreversible reaction for a small excursion of current so that equation (2.1) can be linearised should allow extrapolation as $t^{1/2}$ by the equation

$$\eta = \frac{RT}{nF} \frac{i}{A} \left[\frac{2t^{1/2}}{nF\pi^{1/2}} \left(\frac{1}{C_O^b D_O^{1/2}} + \frac{1}{C_R^b D_R^{1/2}} \right) + \frac{1}{i_0} \right] \tag{2.92}$$

where C_R^b, C_O^b are finite at $t = 0$. However, the method is limited by the double layer charging which dominates the potential charge at short times.

A double current pulse[30] experiment, when R is absent from the bulk, will also characterise a redox process if this is not totally irreversible. A typical profile is shown in Fig. 2.1(d), The relation

$$\frac{t_i}{\tau_d} = \frac{2i_2}{i_1} + \left(\frac{i_2}{i_1} \right)^2 \tag{2.93}$$

is predicted;where t_1, i_1 are the magnitude and height of the first pulse, i_1 the magnitude of the second pulse. t_1 is less than, or equal to τ.

(ii) *Preceding chemical reaction*

Preceding chemical reactions of the type

$$Y \underset{k_{-1}}{\overset{k_1}{\rightleftharpoons}} O \tag{2.94}$$

$$O + ne \rightleftharpoons R$$

give the following[31,32,33] theoretical equation for a single current pulse

$$i\tau^{1/2} = \frac{nFA(\pi D)^{1/2}C^b}{2} - \frac{i\pi^{1/2}\,\text{erf}\,[(k_1 + k_{-1})\tau]^{1/2}}{2\dfrac{k_1}{k_{-1}}(k_1 + k_{-1})^{1/2}} \qquad (2.95)$$

where $C^b = C_Y{}^b + C_O{}^b$ expresses the total concentration in the bulk. Two limiting cases of this have occurred in practice .When

$$[(k_1 + k_{-1})\tau]^{1/2} > 2$$

then

$$i\tau^{1/2} = \frac{nFA(\pi D)^{1/2}C^b}{2} - \sigma i \qquad (2.96)$$

where

$$\sigma = \frac{\pi^{1/2}}{\left(2\dfrac{k_1}{k_{-1}}(k_1 + k_{-1})^{1/2}\right)} \qquad (2.97)$$

It is clear that $i\tau^{1/2}$ plotted against i falls; the gradient gives k_1 and k_{-1} if k is known.

When

$$[(k_1 + k_{-1})\tau]^{1/2} < 01 \qquad (2.98)$$

that is at high currents.

$$i\tau^{1/2} = \frac{nFA(\pi D)^{1/2}C^b}{2\left(1 + \dfrac{k_1}{k_{-1}}\right)} = \frac{nFA(\pi D)^{1/2}C_O{}^b}{2} \qquad (2.99)$$

Equation (2.99) is independent of i and could in principle be used to calculate

$$K = \frac{k_{-1}}{k_1}$$

(iii) *Catalytic reaction*

Catalytic reactions[34] identified as currents in excess of the pure diffusive contribution are observed:-

$$O + ne \rightleftharpoons R$$

$$R + Z \xrightarrow{k} O \qquad (2.100)$$

In the bulk $C_R{}^b = 0$, and Z is not reducible. In a single pulse

$$\left(\frac{\tau}{\tau_d}\right) = \frac{2\gamma}{\pi^{1/2}\,\text{erf}\,(\gamma)} \tag{2.101}$$

where $\gamma = (kC_Z{}^b\tau)^{1/2}$. Experiment and theory can be compared to give k.

(iv) *E.C. reaction*

The E.C. mechanism,[35] written as an anodic reaction,

$$A \overset{k}{\rightleftharpoons} B + ne$$

$$B \overset{k}{\to} Y \tag{2.102}$$

gives, in a double current pulse experiment

$$j\,\text{erf}\,(\theta k\tau_2)^{1/2} = \text{erf}\,(k\tau_2)^{1/2} \tag{2.103}$$

where

$$j = \frac{i_1}{i_1 + i_2};\qquad \theta = \frac{\tau_1 + \tau_2}{\tau_2} \tag{2.104}$$

k can obviously be calculated.

A refinement of the E.C. type[35,36] namely

$$A \rightleftharpoons B + ne \tag{2.105}$$

with

$$B + X \overset{k_1}{\to} L \quad \text{slow}$$

$$L + B \overset{k_2}{\to} A + Y \quad \text{fast}$$

has also been treated under the conditions that C^b is zero in the bulk. Under single pulse conditions, where k_1 is large and the current, i, low so that a stationary concentration of B is attained, the transition time is given by

$$(i\tau^{1/2}) = 2i\tau_d{}^{1/2} - \frac{\pi^{1/2}i}{2(2k)^{1/2}} \tag{2.106}$$

where $k = k_1 C_x$.

The high current conditions means that a stationary concentration of B in not possible and

$$i\tau^{1/2} \to i\tau_d{}^{1/2}.$$

POLAROGRAPHY

A large number of reactions have been studied in the past by this method; comprehensive reviews are available (for example refs. 37, 38). However, only slow processes compared to diffusion can be investigated in any detail. The reason is that measurement is made at or near drop fall ($\simeq 4$ sec) which means that the effective diffusion layer thickness is large. Many excellent accounts are available. In this section some important results will be described briefly.

(i) *Electron transfer*

The simple redox reaction

$$O + ne \rightleftharpoons R$$

can be handled theoretically by solving equations of the form

$$\frac{\partial C_O}{\partial t} = D_O \frac{\partial^2 C_O}{\partial x^2} + \frac{2}{3} \frac{x}{t} \left(\frac{dC_O}{dx} \right)$$

$$\frac{\partial C_R}{\partial t} = D_R \frac{\partial^2 C_R}{\partial x^2} + \frac{2}{3} \frac{x}{t} \left(\frac{dC_R}{dx} \right)$$

$$(2.107)$$

instead of equations (2.38) (2.39) on p. 37. The extra term describes the movement of the mercury surface and its electrolyte layer. The most important result is the Ilkovic equation for the average diffusion current which would be observed at high potential

$$\bar{i}_d = 607n \, C_O{}^b \, D_O{}^{1/2} m^{2/3} t_1{}^{1/6} \tag{2.108}$$

where t_1 is the drop time. Equation (2.108) is calculated for planar diffusion. It is possible to improve the equation by considering the spherical nature of the mercury drop but the correction is of second order importance.

The complete shape of the wave can be analysed as in the section on the rotating disc (p. 83), however, in the case of polarography

$$\delta = (\tfrac{3}{7}\pi D t_1)^{1/2} \tag{2.109}$$

and this can be substituted in equation (2.194) instead of equation (2.195). Refinements are possible if curvature is considered in the treatment and the reader is referred to specialised papers on the detailed analysis.[38a]

(iia) *A first order preceding chemical reaction*

The first order preceding reaction

$$B \underset{k_2}{\overset{k_1}{\rightleftharpoons}} A$$

$$A + e \rightleftharpoons C \tag{2.110}$$

has been solved with sufficient accuracy by the reaction layer model[39,40,41]

$$\frac{\bar{\imath}_l}{\bar{\imath}_d} = \frac{0.886 \, k_2^{1/2}}{1 + 0.886 \left(\dfrac{D_A}{D_B}\right)^{1/2} k_2^{1/2} \, t_1^{1/2} \, K} \tag{2.111}$$

where

$$K = C_A{}^b / C_B{}^b. $$

(iib) *The second order preceding chemical reaction*

$$B \underset{k_2}{\overset{k_1}{\rightleftharpoons}} 2A \quad \text{(a)}$$

$$A + e \rightleftharpoons C \quad \text{(b)} \tag{2.112}$$

has been calculated by Koutecky and Hanus[42] under the conditions

$$C_B = C_B{}^b, \quad C_C = C_C{}^b, \quad C_A{}^b = \left(\frac{C_B{}^b}{K}\right)^{1/2} \tag{2.113}$$

in the bulk of solution. If reaction (2.112a) is predominantly to the left, i.e. in favour of B and

$$\frac{\bar{\imath}_l}{\bar{\imath}_d} = \bar{f}_1(\alpha_1) \tag{2.114}$$

$$\alpha_1 = \left(\frac{k_1 t_1}{(C_B{}^b K)^{1/2}}\right)^{1/2} \tag{2.115}$$

$$K = \frac{k_2}{k_1} = \frac{C_B{}^b}{(C_A{}^b)^2} \tag{2.116}$$

t_1 is the drop time. The function $\bar{f}_1$ is shown in Fig. 2.16, $\bar{\imath}_l$ is the average limiting current; $\bar{\imath}_d$ is the limiting current which would be observed if the current was controlled purely by diffusion, that is with k_1, k_2 fast and $\bar{\imath}_d$

given by the Ilkovic equation. In the original paper[42] a similar graph to Fig. 2.16 is given for instantaneous currents at drop fall.

Dimerisation of C

$$A + e \rightleftharpoons C$$

$$2C \overset{k_2}{\to} \text{products} \tag{2.117}$$

leads to

$$\frac{\bar{i}_l}{\bar{i}_d} = f_2(\gamma_1) \tag{2.118}$$

$$\gamma_1 = \left(\frac{D_C}{D_A} k_2 C_A{}^b \lambda^3 t_1 \right) \tag{2.119}$$

$$\lambda = \exp \left(\frac{nF\eta}{RT} \right) = \left(\frac{C_A}{C_C} \right)_{x=0} \tag{2.120}$$

when the dimerisation is irreversible and fast and C is absent in the bulk. The function $\bar{f}_2$ is defined in Fig. 2.17 from which k_2 and the concentration dependence of $\bar{i}_l/\bar{i}_d$ can be seen. It is clear that in this case the current must always be diffusion controlled at large overpotential.

Catalytic regeneration[42] can be followed by a similar method. The reaction is

$$A + e \rightleftharpoons B$$

$$2B \overset{k}{\to} A + C \tag{2.211}$$

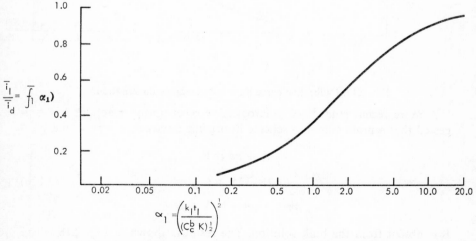

FIG. 2.16. The exact solution for a polarographic second order preceding reaction.[42]

and leads to an exact solution for a rapid reaction, if $C_B{}^b$ is zero,

$$\frac{\bar{i}_l}{\bar{i}_d} = \bar{f}_2(\gamma_1')$$
(2.122)

$$\gamma_1' = \left(\frac{C_A{}^b kt_1'}{(2\lambda + 1)^3}\right)^{1/3}$$
(2.123)

$$\lambda = \exp\left(\frac{nF\eta}{RT}\right) = \left(\frac{C_A}{C_C}\right)_{x=0}$$
(2.124)

f_3 is given by Fig. 2.17 and the relation

$$\bar{f}_3(y) = \bar{f}_2\left[\left(\tfrac{2}{3}\right)^{1/2} y\right].$$
(2.125)

The corresponding situation for a slow reaction is given in ref. 43.

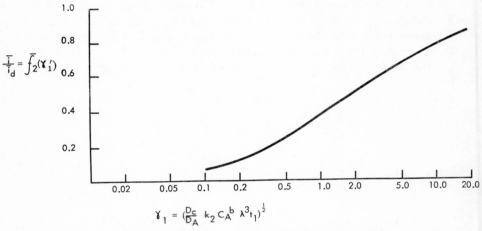

FIG. 2.17. Calibration curve for a polarographic dimerisation.[42]

A more recent study[44] of polarographic regeneration reactions has suggested that a more common scheme in organic electrochemistry would be

$$A + ne \rightleftharpoons B$$

$$2B \overset{k_2}{\to} A + C$$
(2.126)

$$mB \to (m - 1)A + D$$

B is absent from the bulk solution. The result is shown in Fig. 2.18 where $\bar{i}/\bar{i}_d$ is given as a function of $k_2 C_A{}^b \tau$, and as a function of m. k of reaction

(2.121) is $2k_2$ by definition. In the diagram the theory for rapid and slow reactions where the theory can be calculated fairly exactly are joined to make a smooth curve which covers the whole region.

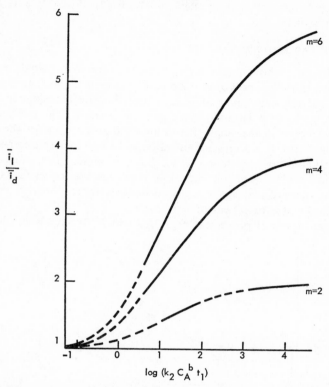

FIG. 2.18. Full line calculated, broken line interpolated. Calibration curve for polarographic catalytic reaction.[44]

Comparison of polarography with other methods

Consider a first order preceding chemical reaction as a suitable example.

$$B \underset{k_2}{\overset{k_1}{\rightleftharpoons}} A$$

$$A + e \rightleftharpoons C \tag{2.127}$$

It can be shown that the upper limit is given by

$$k_2 = \frac{200}{t_1(k_1/k_2)^2} \tag{2.128}$$

where t_1 is the drop time above this value of k_2 the limiting current is indistinguishable from diffusion control by the total concentration. The shortest t_1 is 2 secs. In the potentiostatic, galvanostatic, linear sweep, rotating disc, measurements can be made at much shorter times. Probably a k_2 10^4 times faster can be measured.

SINGLE LINEAR POTENTIAL SWEEP

The fact that the whole potential range can be scanned in a single experiment gives power to the method of linear potential sweep. A substantial advantage over the galvanostatic method is that the charging current can easily be measured in the base electrolyte without the reactant. The method is most easily interpreted in the single sweep mode, the first cycle of the potential profile Fig 2.1(e). It is sometimes advantageous however to observe the current due to continuous potential cycling, for example, when an electroactive intermediate builds up in concentration. However, the theory for continuous cycling to observe stationary state currents is more complex than for a single sweep and is rarely calculated.

Two main types of processes (Figs 2.19 and 2.20) have been identified.

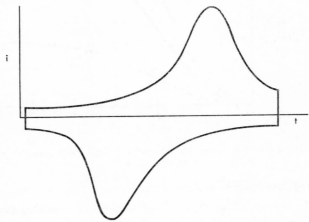

FIG. 2.19. Linear potential sweep curve for a redox reaction (diagrammatic).

Electron transfer with diffusion in the solution

Figure 2.19 shows a typical reaction governed by electrochemical reaction and diffusion in the solution. As the potential advances the current rises until diffusion causes it to drop. The maximum current, i_p, and potential E_p, can be measured as a function of sweep rate and bulk concentration of reactant, to measure the kinetics. The relation of the parameters to one another is the subject of the following sections; information about the kinetics is also contained in the foot ot the wave. It is often possible to obtain a Tafel slope as a function of sweep rate which should be complementary to the behaviour of i_p, E_p. A third source of information is the shape of the wave perhaps measured by the width of the wave at suitable points. The behaviour of the current on reverse sweep as shown in Fig. 2.19 is also obviously important.

Electron transfer with adsorption

A well known surface reaction is shown in Fig 2.20; that of hydrogen deposition on to platinum. The peaks in the region A are due to the build up and removal of a monolayer of hydrogen atoms, which are stable in this potential region and do not react further. The behaviour shown in Fig. 2.20 can be distinguished from that of Fig. 2.19 in that the charge under the peaks is independent of sweep rate and equal to a monolayer. The peak current also has a characteristic dependence on sweep rate ((i) and (viii) below). Much more complicated reactions have been investigated for example the intermediates in methanol oxidation.[45]

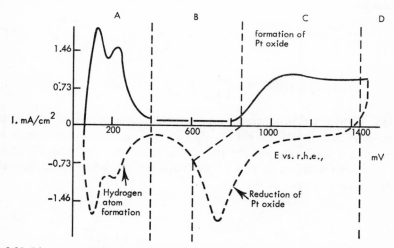

FIG. 2.20. Linear potential sweep experimental curve for a surface process. The deposition of hydrogen on Pt.[56]

(i) Electron transfer

The simple redox reactions[46]

$$O + ne \rightleftharpoons R$$

respond to a single linear sweep as shown in Fig. 2.19. The curve rises sharply as the potential increases, accelerating the reaction until the growing diffusion layer begins to exert control and the current goes down. Other features of the system are noted in the Fig. 2.19. The experimental parameters which characterise a peak are height (i_p), peak potential (E_p) and shape. The shape can be denoted by the width at $i_p/2$; this last test is rarely used. A comprehensive review of the expected results for various reaction schemes is given in the literature.[47] The most important and well known results are for a reversible simple redox reaction at a planar electrode when R is absent in the bulk.

$$E_p = E_{1/2} - 1\cdot109\frac{nF}{RT} \tag{2.129}$$

$$i_p = 0\cdot269An^{3/2}C_O{}^b D_O{}^{1/2}v^{1/2} \tag{2.130}$$

E_p is independent of sweep rate. A completely irreversible charge transfer

$$O + ne \rightarrow R$$

behaves as

$$E_p = E^0 + \frac{RT}{\alpha nF}\left[-0\cdot78 + \ln\frac{k_1}{D_O{}^{1/2}} - \tfrac{1}{2}\ln\left(\frac{\alpha nF}{RT}v\right)\right] \tag{2.131}$$

$$i_p = 0\cdot495\,AnF\left(\frac{\alpha n'F}{RT}\right)^{1/2}D_O{}^{1/2}C_O{}^b v^{1/2} \tag{2.132}$$

where k_1 is the reaction constant at the normal hydrogen reversible potential. i is proportional to $v^{1/2}$ as before. E_p depends now on sweep rate, a fact which is used to distinguish the two cases. In essence, plotting E_p against $\log v$ corrects for diffusion, and a Tafel slope is obtained directly. Rewriting equation (2.131) gives

$$\frac{dE_p}{d\log v} = -\frac{b}{2} \tag{2.133}$$

where b is the Tafel slope.

The reversible and totally irreversible limiting cases can usually be attained in a practical system by adjusting the sweep rate and conditions. Figure 2.21 shows i_p versus $v^{1/2}$ diagrammatically, and, the change from equation (2.130) to equation (2.132).

Nicholson and Shain[47] also treat a number of schemes which were described in the introduction to Chapter 2. They calculate current–time curves for reversible and completely irreversible charge transfer in each case. A synopsis of their results is shown in Fig 2.22. The dependences shown in Fig. 2.22 are quite distinct for different reaction schemes and show clearly why the sweep method is powerful as a diagnostic tool.

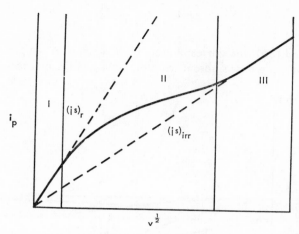

FIG. 2.21. The crossover from completely reversible to irreversible reaction shown by the dependence of i_p on $v^{\frac{1}{2}}$. (I) reversible, (III) irreversible.[46]

(ii) *Preceding chemical reaction*

The preceding chemical reaction[48,49]

$$I \underset{k_b}{\overset{k_f}{\rightleftharpoons}} O \tag{2.134}$$

or

$$I \underset{k_b}{\overset{k_f}{\rightleftharpoons}} 2O \tag{2.135}$$

coupled to the charge transfer reaction

$$O + ne \rightleftharpoons R$$

have been compared by Saveant and Vianello.[49] These authors[48,49] have considered the complete solution when I, O and R are all present in solution. The preceding chemical reaction equation (2.134) is characterised by

$$K = k_f/k_b$$

$$\lambda = \frac{(k_f + k_b)}{v}\left(\frac{RT}{nF}\right). \tag{2.136}$$

A current function ψ is also defined

$$\psi = \frac{i}{nFAD^{1/2}C\left(\dfrac{nFv}{RT}\right)^{1/2}} \tag{2.138}$$

If i_p is inserted for i, the corresponding ψ_p is defined. A simpler situation[47] exists, however, when R is absent from the bulk and the first order preceding

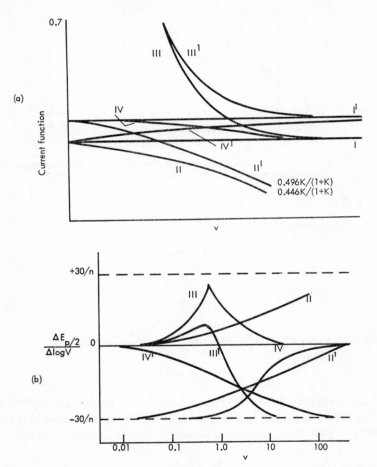

FIG. 2.22. (a) Dependence of i_p on v for various reaction schemes. The y axis is proportional to i_p. (b) Rate of shift of potential with sweep rate. The roman numerals denote the reaction schemes given on p. 29. I′ II′ III′ refer to an irreversible charge transfer. IV′ corresponds to a reversible chemical reaction.[53]

reaction (2.134) precedes discharge. For this case the discharge reaction can either be reversible or irreversible. Figure 2.23 shows that for intermediate values of λ the behaviour of the electrochemical reaction is effectively eliminated by plotting i_p/i_d, that is the ratio of actual to calculated i_d (pp. 62, 63). The nomenclature is given in Fig. 2.23. Similar conclusions hold when O, I, R, are present in solution. The more general case will now be discussed.

For small values of λ the wave is diffusion controlled by the concentration of O. If the total concentration $C_I^b + C_O^b + C_R^b$ is known K, clearly, can be calculated. When λ is high, diffusion will again be observed, this time controlled directly by the total concentration $C_I^b + C_O^b + C_R^b$; K can also be calculated if E^0 is known. Intermediate values of λ show control both by diffusion and the kinetics of the chemical step. These results hold at some fixed value of K. The complete behaviour for all values of K and λ is defined by Fig. 2.24 and Fig. 2.25 for a reversible electrochemical reaction. From Fig. 2.24 at a particular K, λ is defined, Fig. 2.25 shows how the system should behave. In particular in the diffusion controlled region peak height will be proportional to $v^{1/2}$. The peak potential will be independent of sweep rate if the interfacial reaction is reversible. In addition, the peak height will

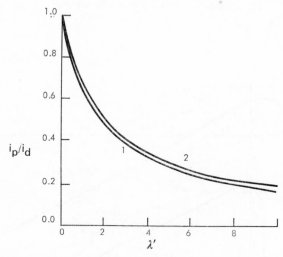

FIG. 2.23. i_p/i_d as a function of λ'. Curve 1 is an irreversible charge transfer,

$$\lambda' = \left(\frac{\alpha n_a Fv}{RT}\right)^{\frac{1}{2}} \frac{1}{K(k_f + k_b)^{\frac{1}{2}}}.$$

Curve 2 is a reversible charge transfer [47]

$$\lambda' = \left(\frac{nFv}{RT}\right)^{\frac{1}{2}} \frac{1}{K(k_f + k_b)^{\frac{1}{2}}}$$

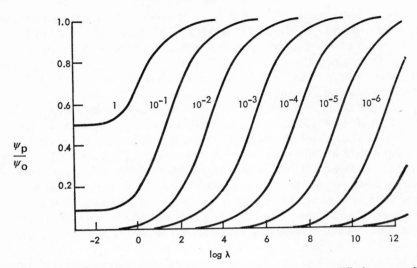

FIG. 2.24. ψ_p/ψ_0 as a function of $\log \lambda$. ψ_0 is the value of ψ_p for a diffusion controlled reversible process. The numbers are values of K. ψ_p is the current at the peak. ψ_0 the value for a diffusion controlled reversible curve.[48]

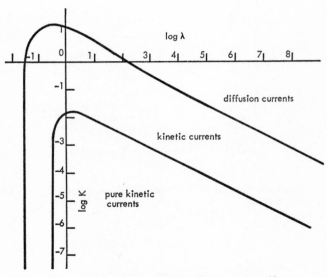

FIG. 2.25. Existence of various current types.[43]

be proportional to concentration. All exactly as discussed in a previous section for a simple redox reaction (see p. 61).

A particularly important case can be observed when K becomes low and λ high. "Pure kinetic currents" are then observed as a limiting value i_p', instead of a peak, having the magnitude

$$i_p' = nFA(C_I + C_O + C_R)D^{1/2}K(k_b)^{1/2}. \tag{2.139}$$

The current–time curve shifts in potential with v, but i_p' itself is independent of v. In practice the sweep rate and concentrations would be adjusted over a wide range to bring the current into a region λ, K could be estimated.

Equation (2.135), on the other hand, shows

$$i_p' = nFA\left(\tfrac{4}{3}\right)^{1/2}D_O^{1/2}C_I^{3/4}K^{3/4}k_b^{1/2} \tag{2.140}$$

in the purely kinetic region and is a means of distinguishing the two reaction schemes. In this case the calculation assumes $C_R^b = 0$. Because two molecules of O are produced in reaction (2.135), when λ is small, the peak height is proportional to C_1^2.

(iii) Catalytic reaction, first order

The catalytic first order reaction

$$O + ne \rightleftharpoons R$$

$$R + Z \overset{k}{\to} O + x \tag{2.141}$$

has been discussed when $C_R^b = O$ in the literature for both reversible and irreversible charge transfer.[47,50] The current–time curves can be described by

$$\lambda = \frac{RT}{nF}\frac{kC_2}{v}. \tag{2.142}$$

At high potentials, for both the reversible and irreversible situation, a limiting current is observed which has the form

$$i_p' = nFAC_O^b(DC_Zk)^{1/2} \tag{2.143}$$

from which k can be estimated.

For a reversible charge transfer, experiment and theory can be compared by the theoretical curve of Fig. 2.26. i_p/i_p' can be measured and plotted against $\log v$. Nicholson and Shain[47] suggest i_p/i_{rev} versus $\lambda^{1/2}$ can be compared with a theoretical curve for a reversible charge transfer. They suggest

$$\frac{i_p}{i_{irr}} \text{ versus } (\lambda_{irr})^{1/2} \tag{2.144}$$

is an appropriate graph to compare with theory in the irreversible case, where

$$\lambda_{irr} = \frac{RT}{nF}\frac{kC_2}{\alpha n_a}$$ (2.145)

(iv) and (vi) *E.C. reaction and dimerisation*
The first order suceeding reaction[47]

$$O + ne \rightleftharpoons R$$

$$R \overset{k_1}{\rightarrow} Z$$ (2.146)

and the dimerisation

$$O + ne \rightleftharpoons R$$

$$R + R \overset{k_2}{\rightarrow} Z$$ (2.147)

are most easily discussed together. It has been shown, as might be expected, with $C_R{}^b = 0$, that a single sweep in the cathodic direction is not a good diagnostic test.[50] Olmstead *et al.*[51] suggest that in a triangular sweep experi-

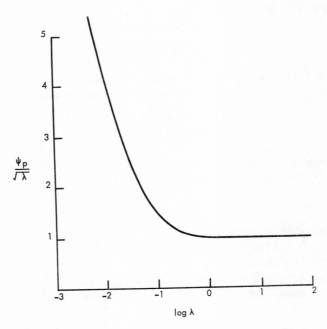

FIG. 2.26. $\psi_p/\lambda^{\frac{1}{2}}$ as a function of log λ. ψ is defined by equation (2.138).[50]

ment at a planar electrode i_a/i_c can be compared with a non-dimensional quantity, ω, where

$$\log \omega = \log k_2 C_O{}^b \tau + 0.034(a\tau - 4) \tag{2.148}$$

and

$$a = \frac{nFv}{RT} \tag{2.149}$$

$$a\tau = \frac{nF}{RT} (E_\lambda - E^0). \tag{2.150}$$

E_λ is the switching potential. E^0 can be estimated from the cathodic peak at high sweep rates where the rate of chemical reaction is small (E^0 is 85% of the cathodic peak when $k_2 C_O{}^b/a \ll 1.5$. A comparison of the first order and second order situation is shown in Fig. 2.27 for $a\tau = 4$.

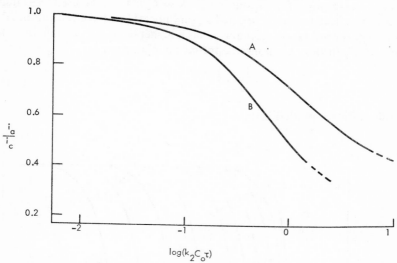

FIG. 2.27. i_a/i_c as a function of $\log (k_2 C_O{}^b \tau)$. A is for a dimerisation ($a\tau = 4$). B is a first order reaction. i_a is the anodic and i_c the cathodic peak height.[51]

(v) E.C.E. reaction and (iii) Catalytic reaction

The E.C.E. mechanism has been discussed in detail.[21,52] In particular when $C_B{}^b = C_C{}^b = 0$ for the scheme

$$A + ne \rightleftharpoons B$$

$$B \underset{k_b}{\overset{k_f}{\rightleftharpoons}} C \tag{2.151}$$

$$C + ne \rightleftharpoons D \qquad \text{E.C.E. mechanism}$$

has been compared to

$$A + ne \rightleftharpoons B \qquad (2.152)$$

$$2B \xrightarrow{k_d} A + P \qquad \text{disproportionation}$$

It is assumed that the charge transfer is completely reversible. The calculated results are formulated in terms of the non-dimensional quantities.

$$\psi_p = \frac{i_p}{0\cdot602n^{3/2}C_A^b D^{1/2}v^{1/2}} \qquad (2.153)$$

$$\lambda = \frac{k_f + k_b}{v}\frac{RT}{nF}. \qquad (2.154)$$

Magstragostino et al.[21] suggest for the E.C.E. mechanism that Fig. 2.28, a plot of ψ_p versus $\log \lambda$ for various $\log K$ ($K = k_d/k_b$), can be used to evaluate the data. The peak height is expected to vary linearly with concentration of A.

The disproportionation mechanism has a higher dependence than linear for i_p versus C_A^b. In this case a more appropriate parameter is

$$\lambda_d = \frac{k_d C_A^b}{v}\frac{RT}{nF} \qquad (2.155)$$

and this is shown plotted in Fig. 2.29.

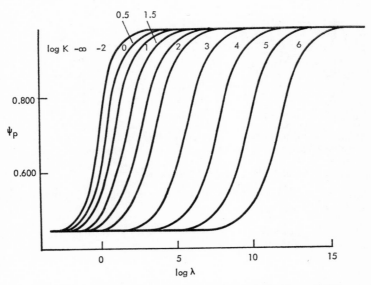

Fig. 2.28. ψ_p variation with $\log \lambda$ for the marked values of K.[21]

The treatment given here for the E.C.E. mechanism is the most general, compared to previous work on this reaction.[53,54]

(viii) *Surface reaction*

Linear sweep was first used to study adsorptive discharge processes by Will and Knorr[55] and was enthusiastically taken up in a qualitative way by many, particularly Breiter,[56] engaged in studies related to fuel cell work. The first theoretical treatment, which was an attempt to reproduce the peaks in region A of Fig. 2.20, was that of Srinivasan and Gileadi,[57] who considered the "Langmuir type" adsorption for $X^- \rightleftharpoons X_{ads} + e$. Only electrochemical reaction and the adsorption isotherm were considered. The effect of diffusion was not investigated. Hale and Greef[58] treated the same mechanism under conditions in which the activation energy is an unspecified function of θ, the coverage, but they confined themselves to two limiting cases; pseudo-equilibrium and negligible back reaction. In the former paper,[57] the equilibrium case gives $E_p = - RT \ln K$, where K is k_1/k_{-1}. Current is proportional to scan rate. The coverages for peak current are 0.5 for equilibrium and 0.63 for the irreversible case. The general case required numerical

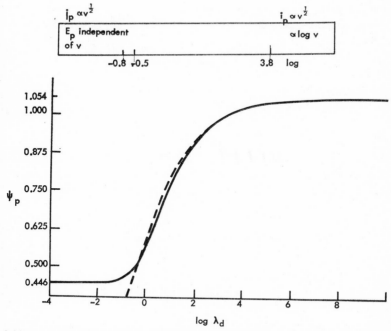

FIG. 2.29. ψ_p as a function of $\log \lambda_d$. The top of the graph shows the two limits of behaviour.[21]

solution. In the latter paper[58] the variation of activation energy with coverage is expressed as a polynomial in integral powers of θ, the first three terms being considered. Depending upon the magnitude of the second and third coefficients, one or two peaks may be obtained (cf. hydrogen adsorption).

A short summary of linear sweep in adsorption discharge[60] has been given by Gileadi and Piersma.[59]

(ix) *Adsorption–desorption*

Rate controlled adsorption–desorption has been described theoretically in a paper by Hulbert and Shain.[61] The adsorption is coupled to a perfectly reversible electrochemical reaction.

$$O + ne \rightleftharpoons R$$

$$R \underset{k_d}{\overset{k_a}{\rightleftharpoons}} R_{ads}$$

(2.156)

The theory is calculated numerically and can deal with a range of values of k_a, k_d. A Langmuir isotherm is assumed. It is probable that the theoretical results depend strongly on this assumption. Given this limitation a typical simulated curve is shown in Fig. 2.30. The adsorption and desorption peaks appear before the diffusion controlled peaks at A and have a characteristic shape and sweep rate dependence from which it is possible in principle[61] to calculate k_a, k_d.

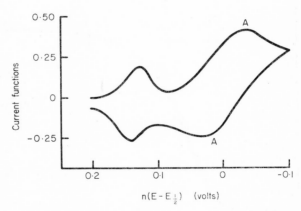

FIG. 2.30. The current function

$$\frac{i}{nFA\ C_O{}^b\ D_O{}^{\frac{1}{2}}\ (nFv/RT)^{\frac{1}{2}}}$$

for the parameters given in the original paper.[61]

BRIDGE METHODS; MEASUREMENT OF IMPEDANCE

The methods described up to this point sample the current, it is a short step to measure the impedance of the electrode by imposing a small a.c. signal on the potential. This amounts to observing the gradient of the current–potential curve at a given potential. The advantage of the a.c. method is that impedances can be measured with high accuracy using bridges methods described in Chapter 1. Small a.c. signals (2 mV amplitude) are used so that the input and output signals are of the sine wave form. a.c. methods are not used as a diagnostic tool to investigate unknown electrochemical mechanisms, however, once a mechanism is known in outline a great deal of fine detail can be revealed.[62,63,64]

(i) *Impedance measured at E_e*

R_s, C_s series circuit. Consider a perturbation of the reaction

$$O + ne \rightleftharpoons R$$

with an a.c. potential about the equilibrium potential. The current response can be calculated directly from the result for a square pulse equation (2.44) or (2.45). The Laplace transform of equation (2.45) is (see equation (2.16)).

$$\bar{\iota}(s) = \frac{I}{s^{1/2}[(sD)^{1/2} + \lambda]} \, . \tag{2.157}$$

It can easily be shown, for example for a linear combination R_s, C_s that the response to a square wave expressed as an impedance is

$$Z(s) = \left(R_s + \frac{1}{sC_s} \right) \tag{2.158}$$

and to a sine wave

$$Z(j\omega) = \left(R_s + \frac{1}{j\omega C_s} \right). \tag{2.159}$$

The more general principle of substituting s by $j\omega$ can be used here. Rearranging equation (2.157) and replacing s by $j\omega$ gives the required stationary equation

$$\bar{Z}(s) = \frac{\bar{\eta}(s)}{\bar{\iota}(s)} \quad \text{and} \quad Z(j\omega) = \frac{[(j\omega D)^{1/2} + \lambda]}{(j\omega)^{1/2}} \frac{\eta(j\omega)}{I(j\omega)} \tag{2.160}$$

Equation (2.160) corresponds to the conditions of equation (2.44) that is when O and R are both present in the bulk solution. Taking the definition of I in equation (2.46) and linearising to evaluate $\eta(j\omega)/I(j\omega)$ gives

$$Z(j\omega) = \frac{RT}{n^2 F^2 i_0} + \frac{RT}{n^2 F^2 i_0} \frac{\lambda}{(j\omega)^{1/2}} \tag{2.161}$$

after a small rearrangement. Equation (2.162) can be interpreted as an impedance by using

$$j^{1/2} = \exp j \frac{\pi}{8} \tag{2.162}$$

to yield directly

$$R_s = \theta + \sigma\omega^{-1/2} = R_D + \sigma\omega^{1-/2} \tag{2.163}$$

where

$$\theta = \frac{1}{Anfi_0} = R_D \tag{2.164}$$

$$\frac{1}{\omega C_s} = \sigma\omega^{-1/2} \tag{2.165}$$

where

$$\sigma = K_d = \frac{1}{An^2 Ff2^{1/2}} \left[\frac{1}{C_O{}^b D_O{}^{1/2}} + \frac{1}{C_R{}^b D_R{}^{1/2}} \right] \tag{2.166}$$

$$\cot \phi = 1 + \frac{\theta\omega^{1/2}}{\sigma}. \tag{2.167}$$

Exactly the same result can be obtained by the traditional method. Starting with equation (1.14), linearising and replacing η by

$$\eta = \eta_0 \cos \omega t \tag{2.168}$$

gives an equation which can be solved in a straight forward manner. Equations (2.163) to (2.166) are illustrated in Fig. 2.31(a) where R_s and C_s have the double layer capacity C_d and solution resistance R_r added. It is obvious that the total impedance is additive and consists of an interfacial part which moves in sympathy with the signal and a diffusional part which can lag behind with a phase angle ϕ. The total impedance of Fig. 2.31(a) will be measured propably as series combination

$$Z^m = R_s{}^m + \frac{1}{\omega C_s{}^m} \, . \tag{2.169}$$

The ohmic resitance R_r can usually be subtracted directly by measuring it at high frequency where R_s, C_s are small. C_{dl} can be measured as a pure capacity in the absence of O (however see later). $(R_s{}^m - R_r)$ and $C_s{}^m$ can be converted to the parallel equivalent $R_p{}^m$, $C_p{}^m$ by using the general conversion

$$C_s = C_p \left(1 + \frac{1}{\omega^2 R_p{}^2 C_p{}^2} \right) \tag{2.170}$$

$$R_s = R_p \left(\frac{1}{1 + \omega^2 R_p{}^2 C_p{}^2} \right) \tag{2.171}$$

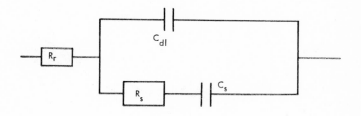

(a)   series analogue

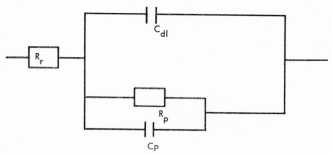

(b)   parallel analogue

FIG. 2.31. Electrical circuit for a redox reaction (a) series analogues, (b) paralled analogue.

C_{dl} can be subtracted from $C_p{}^m$ and parallel elements reconverted to series values C_s, R_s which are plotted according to equations (2.163) to (2.167) see Fig. 2.32(a). It is almost essential to use a programmed desk calculator to do the calculations.

R_p, C_p, *parallel circuit.* There is a lot of advantage to be gained in interpreting the measured values in terms of R_p, C_p the parallel conponents shown in Fig. 2.31(b). Capacitance is then additive. If R_s and C_s of equations (2.163), (2.165) are converted by equations (2.170), (2.171) then

$$R_p = \theta + \frac{\sigma}{\omega^{1/2}}\left(1 + \frac{\sigma\omega^{-1/2}}{\sigma\omega^{-1/2} + \theta}\right) \tag{2.172}$$

$$C_p = \frac{1}{\sigma\omega^{1/2}}\left(\frac{1}{1 + (1 + \theta\omega^{-1/2}\sigma^{-1})^2}\right) \tag{2.173}$$

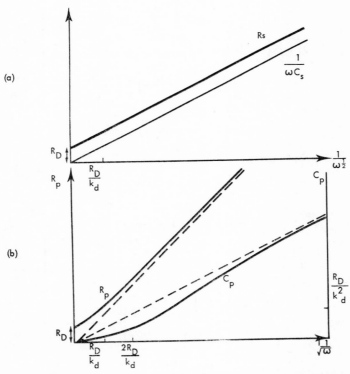

FIG. 2.32. The behaviour of (a) the circuit of Fig. 2.31(a) according to equations (2.163), (2.165); (b) the circuit of Fig. 2.31(b) according to equations (2.172), (2.173).[70]

The behaviour of equations (2.172), (2.173) is shown in Fig. 2.32(b). It is clear from the equations that at high frequencies $R_p \to R_s$, $C_p \to C_s$ hence R_p extrapolates to θ and C_p goes through the origin. As the frequency decreases the dependence of R_p, C_p again becomes linear in $\omega^{-1/2}$. The limiting slope is half that at high frequencies.

The fact that capacitance is now additive allows C_{dl} to be estimated even when a reaction is proceeding. The simplest case occurs when i_0 is large $(\theta \to 0)$ and the reaction is entirely diffusion controlled. A plot[65] of C_p vs. $(\omega R_p)^{-1}$ is linear with unit slope and extrapolates to $C_p = C_{dl}$ at infinite frequency. When however i_0 becomes smaller[66] the method becomes less exact as at low frequencies the limiting value extrapolates to a lower value than C_{dl}. At higher frequencies a curve is observed which still has the correct value, C_{dl}, on the capacity axis as $(\omega R_p)^{-1} \to 0$.

Perhaps a more satisfactory procedure is to consider in more detail what happens when i_0 is large then

$$R_s = \frac{1}{\omega C_s} = \frac{RT}{n^2 F f 2^{1/2}} \left[\frac{1}{C_O^b D_O^{1/2}} + \frac{1}{C_R^b D_R^{1/2}} \right]. \qquad (2.174)$$

If only one component is in low concentration say $C_O \ll C_R$

$$R_s = \frac{1}{\omega C_s} = \frac{\sigma'}{\omega^{1/2} C_O^b} \qquad (2.175)$$

where C_O^b has been separated out from equation (2.173). Converting to parallel curcuit using equations (2.170), (2.171) gives

$$R_p^m = R_p = \frac{2\sigma'}{C_O^b \omega^{1/2}} \qquad (2.176)$$

$$C_p^m = C_{dl} + C_p = C_{dl} + \frac{C_O^b}{2\sigma' \omega^{1/2}}. \qquad (2.177)$$

R_p^m, C_p^m are the measured values when R_r has been removed. R_p^m against $\omega^{-1/2}$ should extrapolate through the origin. C_p should extrapolate to C_{dl}. Both plots have a slope which measures $C_O^b D_O^{1/2}$. In some cases C_{dl} is the same with and without the reacting species O and in other cases not. When C_{dl} is affected the reactant is thought to be specifically adsorbed.

Measurement of θ and hence i_0

The most general use of the a.c. method in the past has been to measure i_0, as the intercept in Fig. 2.32 (the Randles method).

The concentration dependence of i_0 can then be investigated to deduce the kinetics, i_0 is independent of junction potential and is a very useful kinetic quantity. Data on redox reactions and complexes has been obtained by this straightforward method.

Impedance measured at any potential E

In the last section the a.c. potential was applied about E_e. It is more convenient in many cases to apply the a.c. at a fixed potential, using a potentiostat. Similar arguments can be used to the above.

At potentials different from E_e

$$R_s = \theta' + \sigma'' \omega^{-1/2} \tag{2.178}$$

where

$$\theta' = \frac{1}{Anfi_0'} \tag{2.197}$$

$$\frac{1}{\omega C_s} = \sigma'' \omega^{-1/2} \tag{2.180}$$

where

$$\sigma'' = \frac{1}{An^2 Ff 2^{1/2}} \left[\frac{1}{C_O{}^s D_O{}^{1/2}} + \frac{1}{C_R{}^s D_R{}^{1/2}} \right] \tag{2.181}$$

i_0' is the interfacial current alone. At each potential i_0' can be measured by extrapolating the R_s versus $\omega^{-1/2}$ plot to infinite frequency. This procedure is exactly equivalent to other methods (for example, potentiostatic, rotating disc) of extrapolating out diffusion. Log i_0' plotted against E gives a Tafel slope. The concentration dependence and the value of the Tafel slope essentially give the kinetics of the interfacial reaction.

An alternative method of approach is to use the theory to calculate $C_O{}^s$, $C_R{}^s$ and insert these in equation (2.181). The more general approach has been worked out in great detail. However it is much more cumbersome than the experimental method given above.

THE SLUYTERS METHOD

Some results of the more general theory are however of use.

It is possible to measure R_r, C_{dl} in the same experiment by considering the frequency dependence of the complete circuit in Fig. 2.31(a). If the real (resistive) component of the impedance is plotted against the imaginary (capacitative) component, two limiting cases can be derived. A completely reversible reaction would behave as in Fig. 2.33: a completely irreversible reaction as in Fig. 2.34: a moderately reversible reaction gives Fig. 2.35. It would be possible to deduce, from a plot of this kind, $\theta = R_D$ directly. However, Sluyters estimates that only reactions for which $k_{sh} < 10^{-2}$ cm sec^{-1} will be measurable by this technique. He suggests for faster reactions a technique in which the concentrations $C_O{}^b \, C_R{}^b$ are varied. The reader is referred to these references[67,68] for details of this complex plane analysis.

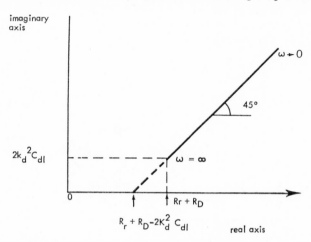

FIG. 2.33. The real and imaginary components of the impedance for a perefctly reversible redox reaction.[64]

(ii) *Preceding chemical reaction*

Consider the reaction

$$Y \underset{k_2}{\overset{k_1}{\rightleftharpoons}} O$$

$$O + ne \rightleftharpoons R$$

the general expression [62,69] for the impedance

$$R_s = \theta + \sigma_R \omega^{-1/2} + \frac{K}{1+K} \sigma_O \omega^{-1/2} + \frac{1}{1+K} \sigma_O \left(\frac{(\omega^2 + k^2)^{1/2} + k}{\omega^2 + k^2} \right)^{1/2}$$

$$(2.182)$$

$$\frac{1}{\omega C_s} = \sigma_R \omega^{-1/2} + \frac{K}{1+K} \sigma_O \omega^{-1/2} + \frac{1}{1+K} \sigma_O \left(\frac{(\omega^2 + k^2)^{1/2} - k}{\omega^2 + k^2} \right)^{1/2}$$

$$k = k_1 + k_2, \qquad K = k_1/k_2 \qquad (2.184)$$

A simplified version of this equation[70,71] which has been shown to occur in practice can be observed when R is large and $\omega \ll k$

$$R_s = \theta + K\sigma_o\omega^{-1/2} + \frac{1}{k^{1/2}(1 + K)}\,\sigma_o2^{1/2} \qquad (2.185)$$

$$\frac{1}{\omega C_s} = K\sigma_o\omega^{-1/2} \qquad (2.186)$$

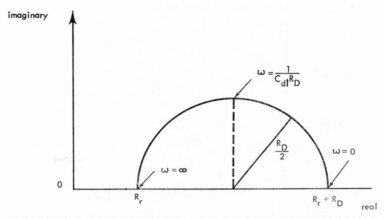

FIG. 2.34. The real and imaginary components of the impedance for a perfectly irreversible redox reaction.[64]

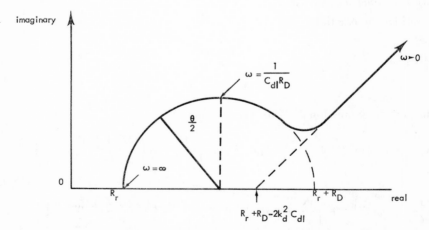

FIG. 2.35. The real and imaginary components of the impedance for a reversible redox reaction. Behaviour intermediate between Figs. 29 and 30.[64]

The form of this equation is shown in Fig. 2.36. According to equation (2.185) and (2.186) the intercept of the extrapolation from low values is

$$\text{intercept} = \theta + \frac{1}{(1 + K)k^{1/2}}\, \sigma_0 2^{1/2}. \tag{2.187}$$

As shown in Fig. 2.36 at high frequencies from equation (2.182), (2.183) the intercept is

$$\text{intercept} = \theta. \tag{2.188}$$

R_s starts to deviate from a linear plot at the point at which

$$\frac{1}{\omega^{1/2}} \sim \frac{1}{k^{1/2}}.$$

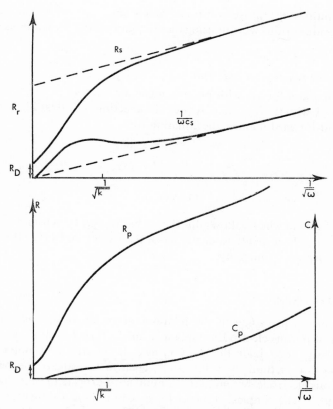

FIG. 2.36. Discharge and diffusion and homogeneous chemical reaction.[70]

The analysis can be taken further, but the basis is essentially contained in equations (2.182), (2.183).

(x) *Adsorption*

Adsorption and diffusion of neutral molecules.

The problem has been solved by Lorenz[72] (however see refs. 73, 74). The equivalent circuit in the general case when the behaviour is controlled by diffusion and a finite rate of adsorption is given by

$$C_p = C_\infty + \frac{\Delta C[(\frac{1}{2}\omega\tau_d)^2 + 1]}{[(\frac{1}{2}\omega\tau_d)^2 + \omega\tau_H]^2 + [(\frac{1}{2}\omega\tau_d)^2 + 1]^2} \tag{2.189}$$

$$\frac{1}{\omega R_p} = \frac{\Delta C[(\frac{1}{2}\omega\tau_d)^2 + \omega\tau_H]}{[(\frac{1}{2}\omega\tau_d)^2 + \omega\tau_H]^2 + [(\frac{1}{2}\tau_d)^2 + 1]^2}. \tag{2.190}$$

The quantities τ_d, τ_H are constants which have the units of a transition time. There are various ways of using these equations and the reader is referred to ref. 74.

(xi) *Preceding reaction and adsorption*

Adsorption and diffusion with preceding chemical reaction.

By analogy with the immediately preceding sections it is possible to derive[75] in a straight-forward manner the equation for

$$Y \underset{k'}{\overset{k}{\rightleftharpoons}} O \tag{2.191}$$

$$O \rightleftharpoons O_{ads}$$

A practical system which follows this theory has not yet been found. Although an attempt has been made to characterise the oxalate–oxalic acid system,[76] and its adsorption at mercury.

POROUS ELECTRODES

It is possible to use some of the relations given above to understand the behaviour of porous electrodes. The most satisfactory way of doing this will probably be to investigate the reaction at a flat electrode by potentiostatic or rotating disc methods as a first step. Then to investigate the behaviour of the porous electrode system itself. A number of expected results have been worked out.[77] The simplest result is for a semi infinite 1-dimensional pore. The reader is referred to ref. 77 for an authoritative account.

ROUGH ELECTRODES

In practise most systems of interest will involve solid metal surfaces. It is essential[77] to know how the a.c. response of the surface itself due to geometric factors will influence the measured impedance. The theory is much less well developed than for porous electrodes. It is expected that roughness effects, for a process which involves diffusion, will be important for times or frequencies at which $Dt/b^2 \ll 1$, b is the height of the surface roughness.

FARADAIC RECTIFICATION

A rather different a.c. method making use of a rectification effect connected with the non-linearity of electrode processes, termed Faradaic rectification.[78] was first examined by Doss and Agarwal.[79] Both the theory and experimental procedures are sophisticated and have not yet been extensively applied. Reference should be made to an extensive paper by Barker[80] for a general analysis to include the effects of adsorption and the presence of a slow chemical step.

A.C. POLAROGRAPHY

A technique which deserves to be mentioned is a.c. polarography. The principle is similar to Section (e) except the amplitude only of the output a.c. signal is measured. The a.c. signal of small amplitude is superimposed on a d.c. potential which is swept through the potential range of interest. The technique has an advantage in that it is automatic. However it suffers in that the theory becomes enormously complicated.[81] It is rarely used for electrode kinetic investigations.

THE ROTATING DISC ELECTRODE

The rotating disc is a powerful tool in electrochemistry.[82,83] The electrode system, shown diagrammatically in Fig. 2.37 is rotated at a strictly controlled and measurable angular velocity, and current or potential control is exercised at the electrode–solution interface according to the desired experimental scheme. At these electrodes there exists a diffusion layer thickness independent of time and radial distance from the axis of revolution, with a well designed electrode. The diffusion layer thickness depends on the angular velocity of revolution in a manner predicted by the theory. In principle, stationary measurements can be made by controlled potential or current. However we would recommend that only the potentiostatic version is used, especially to investigate unknown reactions.

There are other systems which could be used instead of the rotating disc, for example laminar flow through a tubular electrode.[84] The hydrodynamics of a number of systems for example, flow past a sphere, for various reaction schemes have been calculated by Matsuda.[85]

Fig. 2.37

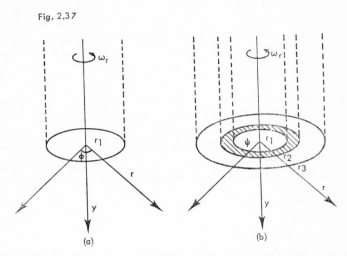

(a)   (b)

FIG. 2.37. Rotating disc assembly (a) simple rotating disc showing usual co-ordinate system; (b) rotating ring-disc.

r_1, radius of the disc;

r_2, radius from the centre of disc to outer edge of the isolating ring;

r_3, radius from centre of disc to outer edge of the metal ring electrode.

Steady state at the rotating disc

The electron transfer reaction

$$O + ne \underset{k_b}{\overset{k_f}{\rightleftharpoons}} R$$

leads to a relation between current and potential at various rotation speeds by the following route: from equation (1.2) in Chapter 1

$$i = nFA(k_f C_O{}^s - k_b C_R{}^s) \tag{2.192}$$

also equation (1.29) shows that the diffusion flux at $x = 0$ is given by

$$i = nFAD_O \left(\frac{C_O{}^b - C_O{}^s}{\delta} \right) = \left(\frac{C_R{}^s - C_R{}^b}{\delta} \right) nFAD_R \tag{2.193}$$

It can easily be shown that in the stationary state the concentration gradient must be linear as given by equation (2.193). Elimination of the surface concentrations between equations (2.193) and (2.192) gives

$$\frac{1}{i} = \frac{1}{nFA(k_f C_O{}^b - k_b C_R{}^b)} + \frac{\delta\left(\dfrac{k_f}{D_O} + \dfrac{k_b}{D_R}\right)}{nFA(k_f C_O{}^b - k_b C_R{}^b)} . \tag{2.194}$$

Substituting the value of the diffusion layer thickness calculated from the hydrodynamics[82,83]

$$\delta = 1\cdot 62\, D^{1/3} \nu^{1/6} \omega_R{}^{-1/2} \tag{2.195}$$

then

$$\frac{1}{i} = \frac{1}{I} + \frac{K}{\omega_R^{1/2}} \tag{2.196}$$

where k is a constant dependent only on potential and I is the current corrected for diffusion. The intercept of $1/i$ versus $1/\omega_R^{1/2}$ gives I which, when measured as a function of potential, gives the Tafel slope. In principle the complete current potential curve could be calculated and compared with experiment, but this is rarely done.

If the electron transfer is perfectly reversible then equation (2.194) becomes, with $k_f \to \infty$, $k_b \to \infty$

$$\frac{1}{i} = \frac{\delta}{AnF} \frac{\left(\dfrac{1}{D_O} - \dfrac{k_b}{k_f D_R}\right)}{\left(C_O{}^b - \dfrac{k_b}{k_f} C_R{}^b\right)} \tag{2.197}$$

k_b/k_f is independent of the kinetics, that is, of k_{sh} and α. From equation (1.9) and (1.10) of Chapter 1

$$k_b/k_f = \exp\left[(E - E^0)fn\right]. \tag{2.198}$$

Exactly the same expression can be derived by substituting in equation (2.192), (2.193) the Nernst equation for the surface concentrations. In contrast to equation (2.196), equation (2.197) has the form

$$\frac{1}{i} = \frac{K'}{\omega_R^{1/2}} \tag{2.199}$$

The graph $1/i$ versus $1/\omega_R^{1/2}$ goes through the origin for all potentials.

Reactions of higher order can also be easily investigated.[86] Consider the totally irreversible reaction

$$pO + ne \xrightarrow{k_f} R. \tag{2.200}$$

If

$$i = nFAk_f(C_O{}^s) \tag{2.201}$$

and as before

$$i = \frac{nFAD_O}{\delta}(C_O{}^b - C_O{}^s)$$

then eliminating $C_O{}^s$ gives

$$i = nFA\,k_f\left(C_O{}^b - \frac{i\delta}{nFAD}\right) \tag{2.203}$$

and by rearrangement

$$\log i = p\log\frac{k_f\delta}{D} + p\log(i_d - i). \tag{2.204}$$

Equation (2.204) can be used in two ways. With a fixed concentration of substance O, the variation of the current at fixed potential on changing δ, i.e. a plot of $\log i$ against $\log(i_d - i)$, has slope p. The calculation effectively measures the order of reaction by the effect of the change in $C_O{}^s$. However, the change in $C_O{}^s$ which can be realised in practice is rather limited. A better method is to use the same plot and vary the bulk, concentration of O.

At high potentials both equations (2.197) and (2.199) predict a maximum current, the diffusion current (see Fig. 1.3) given by

$$i_d = \frac{nFAC_O{}^bD_O}{\delta} \tag{2.205}$$

$$= \frac{nFAC_O{}^b D_O\omega_R{}^{1/2}}{1\cdot62D^{1/3}\nu^{1/6}}. \tag{2.206}$$

The diffusion current is invariably tested in practice according to this equation. If n is known from coulometry D can be calculated.

Equation (2.206) had been extended[87] to electrodes which are partially blocked. This is an important situation which occurs often in practice.

Equation (2.206) becomes

$$i = \frac{nFAC_O{}^b}{\delta + \left| \Sigma A_n \tanh \dfrac{x_n\delta}{r_2{}'} \right|}$$

(2.207)

The value of δ in Fig. 2.38 at which curve b bends, gives an indication, according to the theory, of the extent of coverage. The parameters in equation (2.207) are given in the list of symbols.

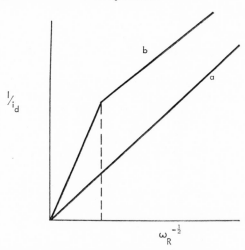

FIG. 2.38. Representation of the effect of an inhomogenous surface on the limiting current at a rotating disc. (a) represents a homogenous electrode. The bend in (b) gives an estimate of the proportion of the surface blocked.

(ii) *Preceding chemical reaction*

For the reaction

$$A \underset{k_{-1}}{\overset{k_1}{\rightleftharpoons}} O$$

(2.208)

$$O + ne \rightleftharpoons R$$

and $k_{-1}/k_1 = K$ the chemical equilibrium constant

$$\frac{1}{\omega_R{}^{1/2}} = \frac{nFD^{2/3}C_O{}^b - C_O{}^s}{1 \cdot 62v^{1/6}} - \frac{KD_A}{1 \cdot 62D^{1/3}v^{1/6}} \frac{i}{D_O\left(\dfrac{k_{-1}}{D_O} + \dfrac{k_1}{D_A}\right)}.$$

(2.209)

Thus a plot of $1/\omega_R^{1/2}$ against i should give a straight line from the slope of which the rate constants (chemical) can be measured. In the above equation

$$D = \frac{k_1 D_O + k_{-1} D_A}{(k_1 + k_{-1})}. \tag{2.210}$$

As an example, Vielstich et al.[88] have studied the dissociation of a number of weak acids.

A preceding reaction of the type

$$B \underset{k_1}{\overset{k_2}{\rightleftharpoons}} 2O \tag{2.211}$$

has been examined theoretically. Under the condition of high reaction rates the limiting current can be calculated[89,90] Hale[91] has used a numerical method to calculate the limiting current at low rates where the reaction k_2 can be ignored. The results are presented in non-dimensional form, $k_1 \delta^2 C_0^b / D$ as a function of $\lambda = i_l / i_d$, where i_d is the expected diffusion current without complication.

(iii) *Catalytic reaction*

For the catalytic reaction

$$O + ne \rightleftharpoons R$$

$$O \underset{k_1}{\overset{k_2}{\rightleftharpoons}} R \tag{2.212}$$

TABLE 2.1.[91]

$\lambda \cdot 10^p$	$p = 2$	$p = 1$	$p = 0$	$p = -1$	$p = -2$
1	0·997	0·964	0·745	0·310	0·0997
2	0·993	0·932	0·611	0·222	0·0704
3	0·989	0·902	0·527	0·182	0·0574
4	0·985	0·874	0·470	0·157	0·0497
5	0·982	0·848	0·427	0·141	0·0444
6	0·978	0·824	0·394	0·129	0·0405
7	0·975	0·802	0·367	0·119	0·0374
8	0·971	0·782	0·345	0·111	0·0350
9	0·968	0·763	0·326	0·105	0·0329

the rates k_1, k_2 can be investigated[91] by measuring the limiting current i_d/i_l $C_R{}^b$, $C_O{}^b$ are the concentrations in bulk. Table 2.1 shows $v(\lambda, 0)$ calculated by solving the appropriate diffusion equations numerically. The experimental $v(\lambda, 0)$ given by

$$v(\lambda, 0) = \frac{i_d}{i_l}(1 + K) - K \qquad (2.213)$$

where $K = k_2/k_1$, allows calculation of λ and of k_1, by the equation

$$\lambda = \frac{(1 + K)\delta^2 k_1}{D}. \qquad (2.214)$$

(iv) *E.C. reaction*

E.C. reactions of the type

$$R \rightleftharpoons T + 2e$$

$$T \xrightarrow{k_1} P \qquad (2.215)$$

have been considered in some detail.[92] In the steady state at the rotating disc the diffusion current of R is independent of k_1. The concentrations at $x = \delta$ are given by $C_T = 0$, $C_R = C_R{}^b$. The current–voltage curve is given by

$$i = \frac{nFAD_R C_R{}^b}{\delta}\left(1 - \frac{1}{\exp\left\{\frac{nF}{RT}(E - E^0 + \alpha_1 + \beta) + 1\right\}}\right) \qquad (2.216)$$

The sign of $E - E_h$ is positive going, for increased positive anodic current. The other quantities in equations (2.175) are defined by

$$\alpha_1 = \frac{RT}{nF}\ln\left\{\left(\frac{k_1}{D_T}\right)^{1/2}\delta \coth\left[\left(\frac{k_1}{D_T}\right)^{1/2}\delta\right]\right\} \qquad (2.217)$$

$$\beta = \frac{RT}{nF}\ln\frac{D_O}{D_R} \qquad (2.218)$$

The half-wave potential as discussed in Chapter 1, where $i/i_d = \frac{1}{2}$ is given by

$$E_{1/2} = E^0 - \alpha_1 - \beta. \qquad (2.219)$$

It is much better to measure k_1 by the potentiostatic pulse or linear sweep experiments described earlier or by the ring-disc, described below. A reversible chemical reaction

$$T \rightleftharpoons P \qquad (2.220)$$

can be treated by the theory on pp. 87, 88 as it is identical in form to a preceding chemical reaction. When another reaction competes for the intermediate T that is

$$R \rightleftharpoons T + 2e$$

$$T \overset{k_1}{\rightarrow} P$$

$$T + C \overset{k_2}{\rightarrow} L \qquad (2.221)$$

$$T + L \overset{fast}{\rightarrow} Q + R$$

where C is in large excess so that k_2 is a pseudo first order rate constant, the current–voltage curve has also been derived. The concentrations at $x = \delta$

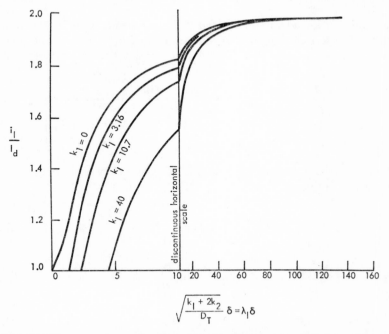

FIG. 2.39. Limiting current as a function of the kinetic parameter $\lambda_1 \delta$, i_l is with coupler, C, and i_d without.[92]

are again $C_R = C_R{}^b$, $C_T = 0$. Perhaps the most important property here is the variation of the limiting current with the parameter

$$\left(\frac{k_1 + 2k_2}{D_T}\right)^{1/2} \delta. \tag{2.222}$$

It k_1 is known, k_2 can be measured[92] (see Fig. 2.39.)

(v) *E.C.E. reaction*

The E.C.E. reaction scheme

$$A + n_1 e \rightleftharpoons B$$

$$B \xrightarrow{k_1} C \tag{2.223}$$

$$C + n_2 e \rightleftharpoons M$$

can be solved for k_1 if the electrochemical rates are diffusion controlled. Figure 2.40 shows a calibration plot of $i/(i_l)_A$ against $\delta^2 k_1/D$

$$\frac{i}{(i_l)_A} = 0.94 \left\{ 1 + \frac{n_2}{n_1} \left[1 - \frac{\left(1 + \dfrac{\delta^2 k_1}{1\cdot 9D}\right)^{1/2}}{\left(1 + \dfrac{\delta^2 k_1}{D}\right)} \right] \right\} \tag{2.224}$$

where, in the bulk, $C_B{}^b = 0$, $C_C{}^b = 0$, $C_A{}^b = C_A{}^b$. Equation (2.224) is derived by solving the convection diffusion equations using Airy functions. In the

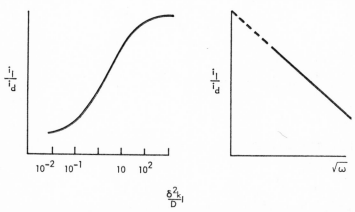

FIG. 2.40. Calibration plot for the E.C.E. mechanism with $n_1 = n_2$.[93]

original paper[93] the author compared equation (2.224) with the earlier, less accurate, calculations of Malchesky *et al.*[94] and Karp.[95]

Reaction scheme (2.223) had been extended[96] to

$$A + e \rightleftharpoons B$$

$$2B \rightarrow C \tag{2.225}$$

$$C + e \rightleftharpoons D$$

The reaction scheme (223) had also been calculated by a numerical method. In both cases the authors suggest a comparison of n_{app} with the function shown in Fig. 2.41 for the first order E.C.E. reaction (2.223). Figure 2.42 gives an equivalent graph for the second order mechanism (2.225). In both cases they have taken into account the equilibria quoted in the figures, this has been commented upon on p. 43.

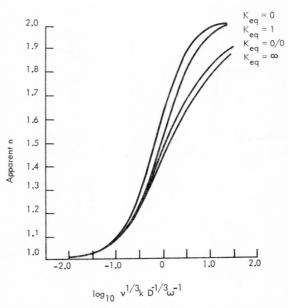

FIG. 2.41. Working curve of n_{app} against $\log v^{1/3} kD^{-1/3} \omega^{-1}$ for the E.C.E. mechanism at a rotating disc.[96]

(vii) *Polymerisation*

The kinetics of electrochemically initiated polymerisation reactions have not been investigated experimentally, however in the last decade interest[97] has grown in the subject. A model scheme for an ionic polymerisation at the rotating disc has been suggested,

$$M + e \rightleftharpoons M_1^* \quad \text{electrode reaction}$$

$$M_2^* + M \xrightarrow{k_p} M_{z+1}^* \quad \text{polymerisation} \qquad (2.226)$$

$$M_{z+1}^* \xrightarrow{k_t} \text{inactive product; termination}$$

The additional condition is that $C_M = C_M^b$, $C_M^{b*} = 0$ at $x = \delta$. Calibration curves, obtained numerically, of $\log [\delta(C_M^b k_p)^{1/2}]$ against i_p/i_M allow experiment and theory to be compared. i_p is the observed limiting current in the presence, and i_m in the absence of polymerisation. Experimental i_p/i_m against $\log \delta$ curves superimposed on Fig. 2.43 gives k_p/k_t and $(k_p)^{1/2}$, for a given C_M^b.

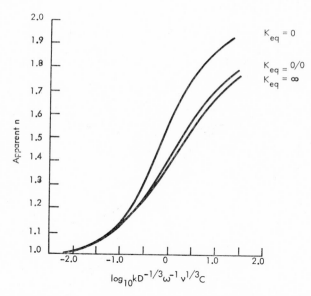

FIG. 2.42. Working curve for a second order E.C.E. mechanism at the rotating disc.[96]

Non-steady state at the rotating disc

The steady state measured does not easily reveal the presence of films of adsorption. It is convenient to carry out transient experiments in the same apparatus. Consider the response of the redox reaction

$$O + ne \rightleftharpoons R$$

to a potentiostatic pulse. At short times compared with the transit time across the diffusion layer, the transient current will have exactly the form predicted on pp. 37, 38 and 39. The complete transient has been calculated by Bruckenstein[98] for the case of reversible metal deposition where O is present in solution and R the solid metal has activity one.

The galvanostatic method has been treated in some detail.[91,99] For the redox reaction it is predicted that transition times will be well defined and longer than those measured in still solution (p. 52). The equations in general are more complex. The analogue of equation (2.90) which describes the transition time is given by the following procedure. y_1/θ^2 is calculated

$$-\frac{y_1}{\theta^2} = \left(\frac{1 \cdot 63 v^{1/6}}{(\omega_R)^{1/2}}\right)^3 \frac{nFAC_O^b}{i\tau^2} \tag{2.227}$$

θ is read off a graph, Fig. 2.44, and is given by

$$\theta = \frac{D^{1/3}\tau}{(1 \cdot 639^{1/6}\omega_R^{-1/2})^2} \tag{2.228}$$

hence D can be calculated. These equations can not be used as equation

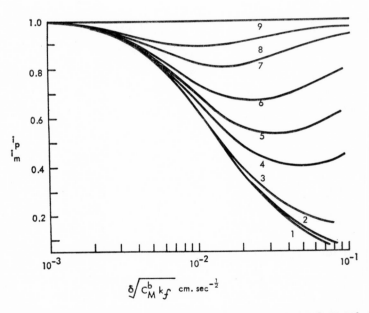

FIG. 2.43. i_p/i_m as a function of δ ($M_\delta k_p$) for values of $M_\delta k_p/k_t$ of (1) 10^7, (2) 10^3, (3) 10^2, (4) 20, (5) 10, (6) 5, (7) 2, (8) 1, (9) 10.[97]

(2.90) to test that the electrochemistry follows a redox reaction. Instead a double pulse method must be used

$$O + ne \rightleftharpoons R \quad \text{current } i \text{ for } t_1$$

$$R \rightleftharpoons O + ne \quad \text{transition time } \tau_2. \tag{2.229}$$

If $C_R{}^b = 0$, then for small i and long times $t_2 \propto 1/\omega_R$. the complete potential time curve for completely reversible redox reaction (compare equation (2.91)) and irreversible reaction (compare equation (2.92)) can also be formulated.

The transition time can be formulated for kinetic and catalytic reactions.[91] Current reversal from the steady state and measurement of the transition time τ_2 will allow calculation of rate constants for first and second order irreversible following reactions when $C_R{}^b = 0$.

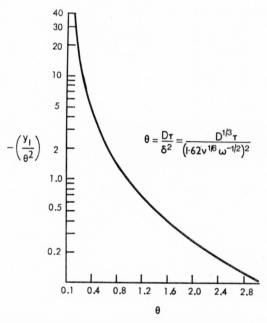

$$\theta = \frac{D\tau}{\delta^2} = \frac{D^{1/3}\tau}{(1\cdot62\nu^{1/6}\omega^{-1/2})^2}$$

FIG. 2.44. Callibration graph for a galvanostatic pulse at the rotating disc.[91,99]

THE ROTATING RING–DISC

The ring–disc electrode shown in Fig. 2.36 is a powerful device for studying intermediates. The amount of intermediates picked up by the ring for purely diffusion and convection of the intermediate, without any chemical reaction across the rotating disc, can be accurately calculated. At a known rotation speed the collection number is only a function of the electrode geometry, that is the radius of the disc and inner and outer radii of the ring. The collection number, N, is defined by

$$i_R = i_D N \tag{2.230}$$

Tables are available[100] for calculating N from known geometry. In practice it is better to calibrate using a simple ion reaction.

Steady state at the ring-disc

When the intermediate experiences an E.C. mechanism of first order[101,102]

$$A + ne \rightleftharpoons B \quad \text{disc}$$

$$B \rightarrow P \qquad \text{in solution} \tag{2.231}$$

$$B \rightleftharpoons A + ne \quad \text{ring}$$

or second order[103,104]

$$A + ne \rightleftharpoons B \quad \text{disc}$$

$$B + C \xrightarrow{k_2} P \qquad \text{in solution} \tag{2.232}$$

$$B \rightleftharpoons A + ne \quad \text{ring}$$

B is absent in the bulk of solution.

In these cases the collection efficiency, N_k, becomes a function of rotation speed and the rate of chemical reaction. For the first order reaction equation (2.231), for thin ring, thin gap electrodes

$$\frac{N}{N_k} = 1 + 1 \cdot 28 \left(\frac{v}{D}\right)^{1/3} \frac{k}{\omega_R} \tag{2.233}$$

when $k < 1$.

A more satisfactory method of calculation is based on a numerical method.[105] The procedure will calculate a callibration curve for any electrode geometry. An example is shown in Fig. 2.45.

In the second order case equation (2.232) the theory can most easily be approximated when the reaction surface is placed at $r = r_2$.

$$i_R = 210 \, nF \pi r_2^2 \, D\omega_R^{3/2} \, v^{-1/2} k^{-1} \tag{2.234}$$

from which k can be calculated. In order to recognise[106,108] when the reaction surface is at r_2, i_R is plotted against i_D for values of ω_R. The linear portion of the graph extrapolates to give i_D ($i_R \to 0$) from which i_D ($r = r_2$) can be calculated by

$$i_D(r = r_2) = \frac{N i_D(i_R \to 0)}{\beta^{2/3}[1 - F(\alpha)]} \tag{2.235}$$

for each ω_R. $F(\alpha)$ is a structural parameter, tabulated in ref. 103 and

$$\alpha = \left(\frac{r_2}{r_1}\right)^3 - 1 \tag{2.236}$$

$$\beta = \left(\frac{r_3}{r_1}\right)^3 - \left(\frac{r_2}{r_1}\right).$$

Given i_D ($r = r_2$) then i_R ($r = r_2$) can be read off the experimental i_D versus i_R graph and k can be estimated from equation (2.234).

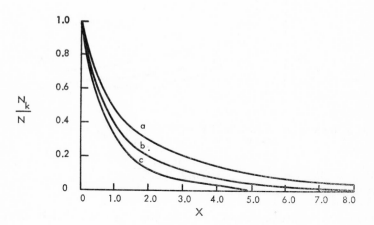

FIG. 2.45. N_k/N as a function of

$$X = \frac{k_1 \, v^{1/3} \, (0 \cdot 51)^{-2/3}}{\omega R \, D^{1/3}}$$

for a first order following reaction[105]

    (a) $r_2/r_1 = 1 \cdot 02$; $r_3/r_2 = 1 \cdot 02$

    (b) $r_2/r_1 = 1 \cdot 04$; $r_3/r_2 = 1 \cdot 28$

    (c) $r_2/r_1 = 1 \cdot 14$; $r_3/r_2 = 1 \cdot 70$

However this method is approximate and holds when $\omega_R/kC_A{}^b$ approaches zero. The complete solution has been obtained numerically.[105] The procedure would be, then, to run the reaction

$$A + ne \rightleftharpoons B$$

completely under diffusion control and measure $N_k = -i/i_D$ as a function of ω_R and the concentration of reaction partner $m = C_C{}^b/C_A{}^b$.

A calibration graph of the type Fig. 2.46 would be necessary for the geometry of the electrode.

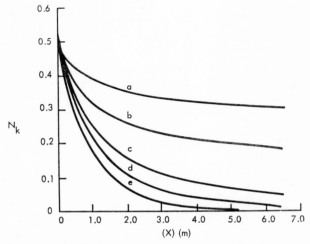

FIG. 2.46. N_k for a second order following reaction $m = C_c{}^b/C_A{}^b$ is (a) 0·1, (b) 0·2, (c) 0·5, (d) 1·0, (e) 10·0, a first order curve.[105]

$$X = \frac{k_2 \, \nu^{1/3} \, (0\cdot51)^{-2/3}}{\omega R \, D^{1/3} \, C_A{}^b}.$$

A more complicated reaction, the reduction of oxygen, has been successfully investigated using a ring disc electrode.[107] Consider the sequence

$$O_2 + 4e \xrightarrow{k_1} H_2O$$

$$O_2 + 2e \underset{k_{-2}}{\overset{k_2}{\rightleftharpoons}} H_2O_2 \tag{2.237}$$

$$H_2O_2 + 2e \xrightarrow{k_3} U_2O$$

O_2 and H_2O_2 are free to diffuse.

The reaction is essentially E.C.E. coupled to a parallel alternative path for electrons. At fixed potential i_R and i_D can be measured at various rotation speeds to give

$$\frac{i_D}{i_R}N = 1 + \frac{2k_1}{k_2} + \frac{\delta_{H_2O_2}}{D_{H_2O_2}}\left[(k_3 - k_{-2}) + (k_3 + k_{-2})(1 + \frac{2k_1}{k_2})\right]. \quad (2.238)$$

It is convenient to rearrange equation (2.238)

$$\frac{i_l - i_D}{i_R} = 1 + \frac{2(k_3 + k_{-2})}{k_2}\frac{D_{O_2}}{D_{H_2O_2}}\frac{\delta_{H_2O_2}}{\delta_{O_2}} + \frac{2}{k_3}\frac{D_{O_2}}{\delta_{O_2}}. \quad (2.239)$$

From these equations k_1 and k_2 can be determined separately, the other parameters k_3, k_1 in combination, if all the constants have a finite value. It is easily possible, from equation (2.238), to derive some limiting cases. If $k_3 = 0$, $k_1 = 0$ then

$$N\frac{i_D}{i_R} = 1 \quad (2.240)$$

which is independent of rotation speed. If $k_1 = 0$ then

$$N\frac{i_D}{i_R} = 1 + \frac{\delta_{H_2O_2}2k_2}{D_{H_2O_2}}. \quad (2.241)$$

Equation (2.241) has fixed intercept and slopes linear in $\omega_R^{-1/2}$ which vary with potential. In general the complete equation (2.238) as a function of $\omega_R^{-1/2}$ has an intercept and slope, both of which vary with potential.

Non-steady state at the ring-disc

The method can also be used in the non-steady state. A potentiostatic or galvanostatic pulse applied to the disc is observed as a current at the ring. There are situations where this is useful, for example when an anodic film and a solution soluble complex are formed in parallel (this is a general feature of some anodic film forming reactions, see Chapter 3). The transit time t', to a potentiostatic pulse is given[109] by

$$\omega_R t' D^{1/3}v^{-1/3}(0.51)^{2/3} = B\left[\log\frac{r_2}{r_1}\right] \quad (2.242)$$

t' is determined as the time from the start of the potentiostatic pulse to the ring current reaching 2% of its final value. B has a value 2.28.

THE THIN LAYER CELL

The basic principle of the method is shown in Fig. 2.47. The working electrode has a thin layer of electrolyte confined near it by an insulating wall, the dimension of the cell is small ($l \sim 4 \times 10^{-3}$ cm). It will be briefly shown that this system has advantages as an experimental method, perhaps the foremost are that the overall number of electrons can be rapidly determined by coulometry and slow chemical reactions can be followed. A major disadvantage is the difficulties of construction, however suitable details for cells operating with mercury and platinum are in the literature.[110]

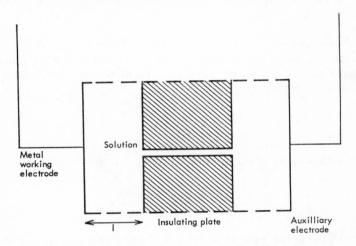

FIG. 2.47. Principle of the mode of operation of thin layer cell.

(i) *Electron transfer*

The simple redox reaction

$$O + ne \rightleftharpoons R$$

has been treated for the linear potential sweep, and galvanostic methods.

(ia) *Linear potential sweep*

If l is small ($\sim 4 \times 10^{-3}$ cm) and rate of potential sweep is small (2 mV sec^{-1}) the reversible reaction has[111]

$$i_p = \frac{n^2 F^2 V v C^b}{4RT} \tag{2.243}$$

where V is the volume of the cell

$$E_p = E^0. \tag{2.224}$$

The wave is symmetrical about E^0 and is the same for the first and subsequent cycles. A totally irreversible, $k_f(E^0) < 10^{-5}$ cm sec^{-1} or $(E_p - E^0) > 0.1$V, wave gives

$$i_p = \alpha n_0 n F v V C_O^b \tag{2.245}$$

$$E_p = E^0 - \frac{2 \cdot 303}{\alpha n_0 F} RT \log \frac{\alpha n_0 F v V}{A R T k_{sh}} \tag{2.246}$$

i_p is linear in v, compare the free diffusion case on p. 62. If an anodic and cathodic peak can be observed in a single cycle, k_{sh} can be estimated without knowing E^0. Totally irreversible stepwise reactions which are separated from one another in potential have also been considered.[111]

(ib) *Galvanostatic method*

The transition time[112] is given by

$$\tau = \frac{n F A l C_O^b}{i} - \frac{l^2}{3D} \tag{2.247}$$

provided $l^2 < \tau D$. Snall value of l further simplifies equation (2.247) to that provided by Faraday's law

$$\tau = \frac{n F A l C_O^b}{i} . \tag{2.248}$$

A reversible reaction should give,

$$E = E^0 + \frac{RT}{nF} \ln \frac{\tau - t}{t} \tag{2.249}$$

it is also claimed that adsorbtion of reacting species can easily be detected.[113] The transition time now becomes

$$\tau = \frac{n F A l C_O^b}{i} + \frac{n F A \Gamma}{i} \tag{2.250}$$

where Γ is the amount of material (moles cm^{-2}) adsorbed. Conditions can be arranged so that contribution from adsorbed reactant is maximised by making l very small ($\sim 10\ \mu$) and the bulk concentration small ($\sim 10^{-3}$ M).

(iv) *E.C. reaction*

E.C. mechanisms[114] in which the chemical rate is small can be investigated by this technique. If a constant current i is sent through the system in the cathodic direction for time t_λ which is interrupted for time t_d, then the transition time for an equal current in the anodic direction is given by

$$\tau = \frac{1}{k} \ln \{[1 + \exp(-kt_d)][1 - \exp(-kt_f)]\} \qquad (2.251)$$

in which k is the first order rate contsant for

$$O + ne \rightleftharpoons R$$
$$\qquad\qquad\qquad\qquad\qquad (2.252)$$
$$R \xrightarrow{k} Z$$

where $C_R = 0$ at $t = 0$.

A thin layer cell with twin electrodes, that is replacing the non-conducting plate by an electrode, has also been used for investigating E.C. mechanisms.[115]

Comment on diffusion (see Table 2.2)

The reader will have noticed a symmtrey in the methods of extrapolation of diffusion. The characteristic quantities are clear from Table 2.2.

TABLE 2.2.   Reaction $O + ne \rightleftharpoons R$.

Methods of extrapolating out diffusion

Method	Extrapolation	Equation
Potentiostatic pulse	$i \propto t^{\frac{1}{2}}$	(2.51)
Galvanostatic pulse	$\eta \propto t^{\frac{1}{2}}$	(2.92)
Potential sweep	$i_p \propto v^{\frac{1}{2}}$	(2.132)
a.c. method	$R_s \propto \omega^{-\frac{1}{2}}$	(2.163)
	$1/w\, C_s \propto \omega^{-\frac{1}{2}}$	(2.165)
Rotating disc	$1/i \propto 1/\omega_R^{\frac{1}{2}}$	(2.196)

CONTROLLED POTENTIAL COULOMETRY

The preceding account has delt exclusively with kinetic schemes which are confined to the diffusion layer. However electrochemists are interested in much slower processes. Examples are to be found in electro-organic synthesis where products accumulate in the bulk of solution. The usual way of investigating reactions of this type is by large-scale electrolysis under potentiostatic conditions.

Two methods can be used to assess the kinetics. One is to measure the current as a function of time and the second is to measure the change in the concentration of the products. If the products are simply distributed it is sufficient to measure the current only, to test various reaction schemes. Suppose the current is measured as a function of time, then the apparent number of electrons (Faradays per mole of substance reacted) can be used as a criterion of the reaction scheme. The current of charge is measured as a function of time. When n has an integral value, then provided the products are known, the interpretation is usually simple. However when n is non integer the reaction path can be complex.

Both the theoretical value of n for a particular reaction scheme and the distribution of products can be calculated as follows.[116,117,118,119,120] It is assumed that the reactant and product diffuse freely in a diffusion layer of negligible volume and reaction of the product or reactant takes place in the bulk of the solution.

The assumption is made that the conditions of stirring are such that uniformity is quickly reached in the bulk of the solution. From the theoretical stand point this has the advantage that the equations have time (t) as a variable and not distance (x). The equations are thus ordinary differential equations not the partial differential with which much of this chapter has been concerned. More complex reaction schemes than hitherto can be calculated.

Consider a simple diffusion controlled reaction

$$O + ne \rightarrow R \tag{2.253}$$

then

$$\frac{dC_O}{dt} = -\frac{p}{V} C_O{}^b = \frac{i}{nFV} \tag{2.254}$$

where p is the rate of mass transfer

$$p = \frac{AD}{\delta} \tag{2.255}$$

and V is the volume of the solution. Integration of equation (2.254) gives

$$C_O{}^b = (C_O{}^b)_i \exp\left(-\frac{p}{V} t\right) \tag{2.256}$$

and

$$i = (i)_i \exp\left(-\frac{p}{V} t\right). \tag{2.257}$$

Therefore if the reaction is completely diffusion controlled, i versus t is linear with a rate constant p/V given by the slope. It the reaction is not completely diffusion controlled, that is controlled by the interfacial rate or by the interfacial rate and diffusion, the same law is obeyed. The rate constant however has a different numerical value.[116] The apparent overall number of electrons can be measured from the current i:-

$$n_{app} = \frac{\int_0^t i\, dt}{FV[(C_O^b)_i - C_O^b]}.$$ (2.258)

For the simple electron transfer described above $n_{app} = n$ and will be integer.

When a preceding, following, or parallel reaction occurs however n_{app} will be non-integer and will depend on p and $(C_O^b)_i$ or the concentrations of other reaction partners. Reaction types which have been given in the literature can be conveniently classified according to whether O or R in

$$O + ne \rightleftharpoons R$$ (2.259)

reacts chemically.[116,117] A typical example is

$$R + O \xrightarrow{k_2} Y.$$ (2.260)

The equations to be solved are

$$\frac{dC_O^b}{dt} = -pC_O^b - k_2 C_R^b C_O^b$$ (2.261)

$$\frac{dC_R^b}{dt} = +pC_O^b - k_2 C_R^b C_O^b$$ (2.262)

with

$$C_R^b = 0 \quad \text{at} \quad t = 0.$$ (2.263)

Solving for C_O^b and evaluating n_{app} at completion of the electrolysis n^0_{app} gives Fig. 2.48. Further diagnostic and quantitative information can also be obtained by current reversal.

Other schemes which have been calculated on this basis are (i) to (vi) quoted at the beginning of the chapter with the more complicated reactions (see Table 1 of ref. 116).

$$O + ne \rightarrow R$$

$$R + Z \xrightarrow{k_1} O$$ (2.264)

$$R + X \xrightarrow{k_2} Y$$

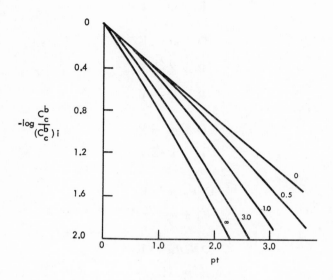

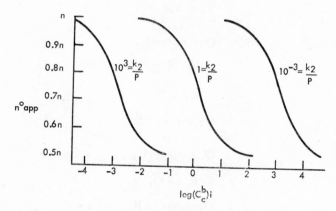

FIG. 2.48. Theoretical coulometric results for the following scheme.[117]

$$C \pm ne \rightarrow R$$
$$R + C \overset{k_2}{\rightarrow} Y$$

(a) n_{app} against $\log C_i$ as a function of $\gamma = k_2/P$

(b) $\log C/(C)_i$ against p_t as a function of $\gamma(C)_i = k_2(C_i^b)_i/P$.

$$O + ne \rightarrow R$$
$$R + O \overset{k}{\rightarrow} T \tag{2.265}$$

$$O + ne \rightarrow R$$
$$R + Z \overset{k_1}{\rightarrow} X \tag{2.266}$$
$$R + O \overset{k_2}{\rightarrow} Y$$

$$O + ne \rightarrow R$$
$$O + X \overset{k}{\rightarrow} X \tag{2.267}$$

$$O + n_1 e \rightarrow R_1$$
$$R_1 + n_2 e \rightarrow R \tag{2.268}$$

$$O + n_1 e \rightarrow R_1$$
$$R_1 \overset{k_2}{\rightarrow} Z \tag{2.269}$$
$$R_1 + n_2 e \rightarrow R$$

$$O + n_1 e \rightarrow R_1$$
$$R_1 \overset{k_2}{\rightarrow} Z$$
$$R_1 \overset{k_1}{\rightarrow} O_2 \tag{2.270}$$
$$O_2 + n_2 e \rightarrow R$$

$$O + n_1 e \rightarrow R_1$$
$$R_1 \overset{k_1}{\rightarrow} O_1$$
$$O_1 + n_2 e \rightarrow R \tag{2.271}$$
$$2R_1 \overset{k_2}{\rightarrow} X$$

Karp and Meites[121] have taken into account the simultaneous reaction in the diffusion layer for the E.C.E. mechanism. They conclude that if the intermediate is detectable in solution then probably the theory given above is accurate and it is not necessary to correct for a small amount of reaction in the diffusion layer.

SYNTHESIS AT CONSTANT POTENTIAL

The last section was particularly weighted towards measuring n_{app}. However the theory can be extended to concentrate on measuring the concentration of intermediates with time. Kinetic data and optimisation of synthesis can both be calculated. An example which has been worked out in some detail[122]

is the E.C.E. mechanism (v)

$$O + ne \rightleftharpoons R$$

$$R \overset{k}{\rightarrow} P$$

$$P + ne \rightleftharpoons D$$

If it is assumed the the reaction is run under completely diffusion controlled conditions the concentration build-up can be predicted by equations of the type (2.261), (2.262). The result is shown in Figs. 2.49 and 2.50. The final analysable concentration of P is $C_P^b + C_R^b$, where C_P^b and C_R^b are the instantaneous concentrations at the end of electrolysis. From graphs of this type it is possible to work out the best conditions for synthesis, or to derive the value of k. It can be shown that for $p/V = 10^{-4} \sec^{-1}$, a typical experimental value in a preparative rotating disc cell, values of k_2 between 10^{-3}–$10^{-5} \sec^{-1}$ can be measured by this method. Other reaction schemes can easily be calculated.

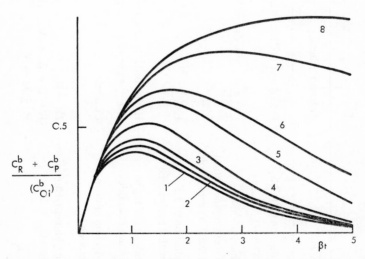

FIG. 2.49. Theoretical final concentration of intermediate P which builds up in solution for the E.C.E. mechanism.[122]

TABLE 2.3

System	Reaction Scheme	Method	Reference
Reduction $Cd(CN)_4^{2-}$ at mercury	Preceding $Cd(CN)_4^{2-} \rightleftharpoons Cd(CN)_3^{-} + CN^{-}$	Potential pulse	17
Dimerisation of Triphenylamine Cation Radical (in aceto nitrile)	E.C.E.	Rotating disc (platinum)	96
Reduction of 3 3' dimethyl azobenzene	E.C.	Linear potential sweep	129
Oxidation of Hg in presence of cyanide	Higher order reaction of the type $niO + ne \rightarrow qR$	Linear potential sweep	130
Reduction of nitro compounds	Catalytic	Polarography	44
Benzyldimethyl anilinium bromide	$2R \rightarrow R_2$ $R + 2 - R' + A$ $R' \pm ne \rightarrow Y$	Coulometry at controlled potential	119

System	Reaction Scheme	Method	Reference
Reduction of azobenzene	E.C. mechanism	Potentiostatic double pulse	131, 132, 129, 22
Reduction of organic acids (acetic acid)	Preceding chemical reaction	Rotating disc	88
Reduction of organic acids	Preceding chemical reaction	Polarography	39, 40
Reduction of (a) p nitrosophenol (b) p nitrophenol	E.C.E.	Single potentiostatic pulse	23
Oxidation of substituted triphenylamines in CH_3CN	E.D.E.	Rotating d sc	
Reaction of As(iii) with I_3^-.	E.C. (second order)	Rorating disc	109
Methyl blueleucomethylene blue	$0 + ne \underset{k_d}{\overset{k_a}{\rightleftharpoons}} R \rightleftharpoons Rads$	Linear sweep	61

Experimental Results

For a detailed discussion of the experimental details the reader must consult the original literature. Fairly complete reviews of the general aspects of the oxidation of organic compounds,[123] and the reduction of organic compounds [124] appear in the literature. Inorganic couples are reviewed inVetters book.[125] Fuel cell reactions[126,127] and batteries[128] are treated in recent books. The experimental data relevant to this chapter are drawn together in Table 2.3 where some systems for which the chemistry and electrochemistry is now fairly well known and have been investigated by the methods described above.

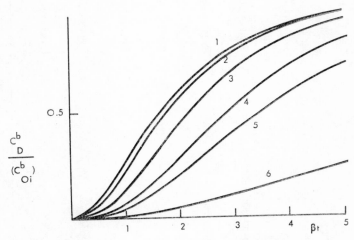

Fig. 2.50. Concentration of species D as a function of time.[122]

References

1. Brown, O. R. and Harrison, J. A. *J. electroanal. Chem.* **21**, 387 (1969).

2. Weinberg, N. L. and Weinberg, H. R. *Chem. Revs.* **68**, 449 (1968).

3. Breiter, M. W. "Electrochemical Processes in Fuel Cells", Springer Verlag, Berlin (1969).

4. Bockris, J. O'M. and Srinivasan, S. "Fuel Cells", McGraw-Hill, New York (1969).

5. Vielstich, W. "Fuel Cells", John Wiley, New York (1970).

6. Meites, L. "Polarographic Techniques", Interscience, New York (1965).

7. Heyrowsky, J. "The Principles of Polarography", Academic Press, London and New York (1966).

8. Randles, J. E. B. *In* "Polarography", (Ed., G. W. C. Milner), Longmans (1957).

9. Churchill, R. V. "Modern Operational Mathematics in Engineering", McGraw-Hill, New York (1944).
10. Jaeger, J. C. "Laplace Transformation", Methuen, London (1961).
11. Olmstead, M. L. and Nicholson, R. S. *J. electroanal. Chem.* **16**, 145 (1968).
12. Oldham, K. B. *Anal. Chem.* **41**, 1904 (1969).
13. Booman, G. L. and Pence, D. T. *Anal. Chem.*, **37**, 1367 (1965).
14. Feldberg, S. W. "Electroanalytical Chemistry", Vol. 3 p. 199, (Ed., A. J. Bard), Dekker, New York (1969).
15. Gerischer, H. and Vielstich, W. *Z. phys. Chem. NF* **3**, 16 (1955).
16. Oldham, K. B. and Osteryoung, R. A. *J. electroanal. Chem.* **11**, 397 (1966).
17. Gerischer, H. *Z. Elek.* **64**, 29 (1960).
18. Koutecky, J. and Brdička, R. *Coll. czech. chem. Comm.* **12**, 337 (1947).
19. Christie, J. H. *J. elecrtoanal. Chem.* **13**, 79 (1967).
20. Feldberg, S. W. *J. phys. chem.* **73**, 1238 (1969).
21. Magstragostino, M., Nadjo, L. and Saveant, J. *Electrochimica Acta.* **13**, 721 (1968).
22. Schwarz, W. M. and Shain, I. *J. phys. Chem.* **69**, 30 (1965).
23. Alberts, G. S. and Shain, I. *Anal. Chem.* **35**, 1859 (1963).
24. Hermann, H. B. and Blount, H. N. *J. phys. Chem.* **73**, 1406 (1969).
25. Hawley, M. D. and Feldberg, S. W. *J. phys. Chem.* **70**, 3459 (1966).
26. Hale, J. M. *J. electroanal. Chem.* **8**, 181 (1964).
27. Olmstead, M. L. and Nicholson, R. S. *Anal. Chem.* **41**, 851 (1969).
28. Koopmann, R. *Ber. Bunsenges.* **72**, 32 (1968).
29. Gierst, L. E. *Z. Elek.* **59**, 784 (1955).
30. Berzins, T. and Delahay, P. *J. amer. chem. Soc.* **75**, 4205 (1953).
31. Gierst, L. and Juliard, A. *J. phys. Chem.* **57**, 701 (1953).
32. Delahay, P. and Berzins, T. *J. amer. chem. Soc.* **75**, 2486, 4205 (1953).
33. Gerischer, H. *Z. phys. Chem.* **2**, 80 (1954).
34. Delahay, P., Mattax, C. C. and Berzins, T. *J. amer. chem. Soc.* **76**, 5319 (1954).
35. Jaenicke, W. and Hoffmann, H. *Z. Elek.* **66**, 803 (1962).
36. Fischer, O. and Dracka, O. *Coll. czech. chem. Comm.* **24**, 3046 (1959).
37. Kuta, J. "Principles of Polarography", (Ed., J. Heyrovsky), Academic Press, London and New York (1966).
38(a). Mairanovskii, S. G. "Galalytic and Kinetic Waves in Polarography", Plenum Press, New York (1968).
38(b). Milner, G. W. C. "Polarography", Longmans, London (1957).
39. Koryta, J. *Z. Elek.* **64**, 23 (1960).
40. Brdcka, R. *Z. Elek.* **64**, 16 (1960).
41. Koryta, J. *In* "Advances in Electrochemistry and Electrochemical Engineering," Vol. 6 (Ed., P. Delahay), Interscience, New York (1967).
42. Koutecky, J. and Hanus, V. *Coll. czech. chem. Comm.* **20**, 124 (1955).
43. Koutecky, J. and Koryta, J. *Coll. czech. chem. Comm.* **19**, 845 (1954).
44. Kastening, B. *J. electroanal. Chem.* **24**, 417 (1970).
45. Breiter, M. W. *Disc. Far. Soc.* April (1968).
46. Matsuda, M. and Ayabe, Y. *Z. Elek.* **59**, 494 (1955).
47. Nicholson, R. S. and Shain, I. *Anal. Chem.* **36**, 706 (1964).
48. Saveant, J. M. and Vianello, E. *Electrochim. Acta* **8**, 905 (1963).
49. Saveant, J. M. and Vianello, E. *Electrochim. Acta* **12**, 1545 (1967).
50. Saveant, J. M. and Vianello, E. *Electrochim. Acta* **10**, 905 (1965).

51. Olmstead, M. L., Hamilton, R. G. and Nicholson, R. S. *Anal. Chem.* **41**, 260 (1969).
52. Olmstead, M. L. and Nicholson, R. S. *Anal. Chem.* **41**, 863 (1969).
53. Nicholson, R. S. and Shain, I. *Anal. Chem.* **37**,   (1965).
54. Saveant, J. M. *Electrochim. Acta* **12**, 753 (1967).
55. Will, F. G. and Knorr, C. A. *Z. Elek.* **64**, 258 (1960).
56. Breiter, M. W. *Electrochim. Acta* **8**, 925 (1963).
57. Srinivasan, S. and Gileadi, E. *Electrochim. Acta* **11**, 321 (1966).
58. Hale, J. M., Greef, R. *Electrochim. Acta* **12**, 1409 (1967).
59. Gileadi, E. and Piersma, B. J. "Modern Aspects of Electrochemistry", Vol. 4 Plenum Press, New York (1966).
60. Gileadi, E. "Electrosorption", Plenum Press, New York (1967).
61. Hulbert, M. H. and Shain, I. *Anal. Chem.* **42**, 162 (1970).
62. Sluyters-Rehbach, M. and Sluyters, J. H. *In* "Electroanalytical Chemistry" Vol. 4 (ed., A. J. Bard), Dekker, New York (1960).
63. Delahay, P. "New Instrumental Methods in Electrochemistry", Interscience, New York (1954).
64. Sluyters, J. H. *Rec. Trav. Chim.* **79**, 1092 (1960).
65. Sluyters, J. H. and Ooman, J. I. C. *Rec. Trav. Chim.* **79**, 1101 (1960).
66. Rehbach, M. and Sluyters, J. H. *Rec. Trav. Chim.* **80**, 469 (1961).
67. Rehbach, M. and Sluyters, J. H. *Rec. Trav. Chim.* **81**, 301 (1962).
68. Sluyters-Rehbach, M. and Sluyters, J. H. *Rec. Trav. Chim.* **82**, 525, 535 (1963).
69. Smith, D. E. *In* "Electroanalytical Chemistry", (Ed., A. J. Bard) Dekker, New York (1966).
70. Gerischer, H. *Z. phys. Chem.* **198**, 286 (1951).
71. Aylward, G. H., Hayes, J. W. and Tamamushi, R. *In* "Electrochemistry, Proc. 1st Australian Conf. 1963", (Eds., J. A. Friend, F. Gutman), Pergamon Press, London (1964).
72. Lorenz, W. *Z. elektrochem.* **62**, 192 (1958).
73. Parsons, R. *In* "Advances in Electrochemistry and Electrochemical Engineering", Vol. 1, (Ed., P. Delahay), Interscience, New York (1961).
74. Armstrong, R. D., Race, W. P. and Thirsk, H. R. *J. electroanal. Chem.* **16**, 517 (1968).
75. Armstrong, R. D. *J. electroanal. Chem.* **22**, 49 (1969).
76. Race, W. P. *J. electroanal. Chem.* **24**, 315 (1970).
77. de Levie, R. *In* "Advances in Electrochemistry and Electrochemical Engineering", Vol. 6, (Eds., P. Delahay and C. W. Tobias), Interscience, New York (1967).
78. Oldham, K. B. *Trans. Far. Soc.* **53**, 50 (1957).
79. Doss, K. and Agarwal, H. *Proc. Ind. Acad. Sci.* **354**, 45 (1952).
80. Barker, G. C. Trans. Symp. Electrode Proc., Philadelphia, 1959, John Wiley, New York (1961).
81. Smith, D. E. "Electroanalytical Chemistry", Vol. 1 Edward Arnold, London (1966).
82. Levich, V. G. "Physicochemical Hydrodynamics", Prentice-Hall, New York (1962).
83. Riddiford, E. C. "Advances in Electrochemistry and Electrochemical Engineering", Vol. 4, Interscience, New York (1966).
84. Matsuda, H. *J. electroanal. Chem.* **15**, 325 (1967).
85. Matsuda, H. *J. electroanal. Chem.* **25**, 461 (1970).

86. Nekrasov, L. N. and Filinovsky, V. Yu. C.I.T.C.E. Meeting, Strasbourg, p. 119 (1969).
87. Scheller, F., Landsberg, R. and Wolf, H. *Electrochemica Acta* **15**, 525 (1970).
88. Jahn, O. and Vielstich, W. *In* "Advances in Polarography", Vol. 1, (ed., I. S. Longmuir), Pergamon Press, New York (1960).
89. Koutecky, J. and Levich, V. G. *Zh. fiz. Khim.* **32**, 1565 (1958).
90. Levich, V. G. *Zh. fiz. Khim.* **32**, 352 (1958).
91. Hale, J. M. *J. electroanal. Chem.* **8**, 332 (1964).
92. Tong, L. K. J., Liang, K. and Ruby, W. R. *J. electroanal. Chem.* **13**, 245 (1967).
93. Filinovsky, V. Yu. *Sov. J. Electrochem.* **5**, 590 (1969).
94. Malachesky, P. A., Marcoux, L. S. and Adams, R. N. *J. phys. Chem.* **70**, 4068 (1966).
95. Karp, S. *J. phys. Chem.* **72**, 1082 (1968).
96. Marcoux, L. S., Adams, R. N. and Feldberg, S. W. *J. phys. Chem.* **73**, 2611 (1969).
97. Gray, D. and Harrison, J. A. *J. electroanal. Chem.* **24**, 187 (1970).
98. Bruckenstein, S. and Prager, S. *Anal. Chem.* **39**, 1161 (1967).
99. Hale, J. M. *J. electroanal. Chem.* **6**, 187 (1963).
100. Albery, W. J. and Bruckenstein, S. *Trans. Far. Soc.* **62**, 1920 (1966).
101. Albery, W. J. and Bruckenstein, S. *Trans. Far. Soc.* **62**, 1946 (1966).
102. Albery, W. J., Hitchman, M. L. and Ulstrup, J. *Trans. Far. Soc.* **64**, 2831 (1968).
103. Albery, W. J. and Bruckenstein, S. *Trans. Far. Soc.* **62**, 2584 (1966).
104. Albery, W. J., Hitchman, M. L. and Ulstrup, J. *Trans. Far. Soc.* **65**, 1101 (1969).
105. Prater, K. B. and Bard, A. J. *J. electrochem. Soc.* **117**, 207, 335 (1970).
106. Albery, W. J., Bruckenstein, S. and Johnson, D. C. *Trans. Far. Soc.* **62**, 1938 (1966).
107. Filinovsky, V. Yu. C.I.T.C.E., Strasbourg, Abstracts, p. 38 (1969).
108. Bruckenstein, S. and Feldman, G. *J. electroanal. Chem.* **9**, 395 (1965).
109. Johnson, D. C. and Bruckenstein, S. *J. amer. chem. Soc.* **90**, 6592 (1968).
110. Hubbard, A. T. and Anson, F. C. *electroanal. Chem.* Vol. 4, (Ed., A. J. Bard), Dekker (1970).
111. Hubbard, A. T. *J. electroanal. Chem.* **22**, 165 (1969).
112. Christenson, C. A. and Anson, F. C. *Anal. Chem.* **35**, 205 (1963).
113. Hubbard, A. T. and Anson, F. C. *J. electroanal. Chem.* **9**, 163 (1965).
114. Christenson, C. A. and Anson, F. C. *Anal. Chem.* **36**, 495 (1964).
115. McDuffie, B., Anderson, L. B. and Reilley, C. N. *Anal. Chem.* **38**, 883 (1966).
116. Bard, A. J. and Santhanam, K. S. V. Electroanalytical Chemistry, Vol. 4, (Ed., A. J. Bard), Dekker (1970).
117. Geske, D. H. and Bard, A. J. *J. phys. Chem.* **63**, 1057 (1959).
118. Bard, A. J. *Anal. Chem.* **40**, 64R (1968).
119. Bard, A. J. and Mayell, J. S. *J. phys. Chem.* **66**, 2173 (1962).
120. Bard, A. J. and Solon, E. *J. phys. Chem.* **67**, 2326 (1967).
121. Karp, S. and Meites, L. *J. electroanal. Chem.* **17**, 253 (1968).
122. Shoesmith, D. W. and Harrison, J. A. *J. electroanal. Chem.* **28**, 301 (1970).
123. Weinberg, N. L. and Weinberg, H. R. *Chem. Rev.* 449 (1969).
124. Brown, O. R. and Harrison, J. A. *J. electroanal. Chem.* **21**, 387 (1969).
125. Vetter, K. J. "Electrochemical Kinetics", Academic Press, London and New York (1967).

126. Srinivasan, S. and Bockris, J. O'M. "Fuel Cells", McGraw-Hill, New York (1969).
127. Breiter, M. W. "Electrochemical Processes in Fuel Cells", Springer, Verlag, Berlin (1969).
128. Jasinski, R. "High Energy Density Batteries", Plenum Press, New York (1967).
129. Lundquist, Jr. J. T. and Nicholson, R. S. *J. electroanal. Chem.* **16,** 445 (1968).
130. Shuman, M. S. *Anal. Chem.* **41,** 142 (1969).
131. Oglesby, D. M., Johnson, J. D. and Reilley, C. N. *Anal. Chem.* **38,** 385 (1966).
132. Shwarz, W. M. and Shain, I. *J. phys. Chem.* **70,** 845 (1966).

Chapter 3

Electrocrystallisation

The problem of the kinetics of electrocrystallisation reactions has had a long history. In the case of metal deposition Erdey–Gruz and Volmer[1] in 1931 suggested that two dimensional nucleation and growth would be the rate determining step. However, no evidence could be obtained for this mechanism. It was then assumed until fairly recently, without any convincing evidence, that growth was similar to the gas phase. Here it was thought that the large number of screw dislocations would control the growth by a surface diffusion mechanism. Recently it has been found experimentally that a two-dimensional nucleation and growth mechanism applied when the number of dislocations is drastically reduced. This leads to the possibility for the first time of investigating the formation of metal or non-metal deposits in an exact way.

INTRODUCTION

In this chapter, metal deposition[2] and anodic film formation (corrosion, passivation, primary and secondary batteries) will be considered together as a phase change. It will be convenient to consider the potentiostatic and galvanostatic response separately for the nucleation and growth model. A brief description of the salient features of the adatom model will then be given. In sections a(1), a(2) the growth of two-dimensional layers on perfect substrates will be discussed in some detail as this must be the fundamental unit which is responsible for lattice building.

NUCLEATION AND GROWTH MODEL (WITHOUT EDGE EFFECTS)

POTENTIAL STEP METHOD

(i) *Without diffusion*

It will be assumed that when a pulse is started nuclei form as discrete centres and grow.[3] The nuclei form either instantaneously or progressively with

115

time. The nuclei do not interact with the boundary. Another feature of the system is that nuclei, on growing, can interact with one another. All these features can be incorporated into a theoretically predicted current time response which can and has been extensively compared with experiment.

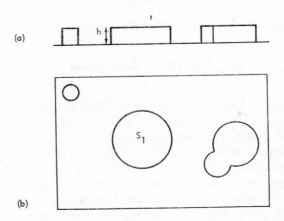

FIG. 3.1. Formation of a thin layer of fixed height on a surface by the growth of two-dimensional nuclei; (a) side view, (b) plan.

Suppose that the situation of Fig. 3.1 is considered, firstly for a single two dimensional nucleus. The growth round the periphery will be controlled by an electrochemical rate of the form equation (1.4) of Chapter 1. In a potentiostatic experiment this will be constant k (moles^{-2} sec^{-1}) say. The current at any instant will be given by

$$i = nF\, kS \tag{3.1}$$

$$nF\, kS = \frac{\rho}{M} nF \left(\frac{dV}{dr}\right)\left(\frac{dr}{dt}\right) \tag{3.2}$$

where S is the area on which material is deposited and V is the volume. Once S and dV/dr are known by taking a particular geometry then integration of equation (3.2) gives r as a function of t. The resulting value of S as a function of time in equation (3.1) gives the result. As an example, Fig. 3.1, for deposition round the edge $S = 2\pi rh$, hence

$$i = nF\, 2\pi h k^2 \frac{M}{\rho} t \tag{3.3}$$

M is the molecular weight, ρ the density of the deposit and h the height of the nucleus, when N_0 nuclei are formed initially equation (3.3) becomes

$$i = nF N_0 \, 2\pi h k^2 \frac{M}{\rho} t. \tag{3.4}$$

A summary of expressions for other geometries which can be calculated in this straight forward manner are shown in Table 3.1.

The extension of equation (3.4) to progressive nucleation can be calculated given that the number of nuclei at any time by

$$N = N_0 \left(1 - (\exp - At)\right). \tag{3.5}$$

The processes of nucleation and growth occur simultaneously and the resulting current is given by

$$i = \int_0^t i(u) \left(\frac{dN}{dt}\right)_{t-u} du \tag{3.6}$$

u is the age of a nucleus. If $i(u)$ is represented by equation (3.3) with t replaced by u, and dN/dt by

$$\frac{dN}{dt} = N_0 \, A \exp\left(-At\right) \tag{3.7}$$

or in particular simply by the approximation

$$\frac{dN}{dt} = N_0 \, A = A \tag{3.8}$$

if N_0 is included in the definition of A; i as a function of t can be estimated. The result for the two-dimensional model of Fig. 3.1 is

$$i = \frac{nF \, \pi M h A \, k^2 \, t^2}{\rho} \tag{3.9}$$

which is to be compared with equation (3.4) for instaneous nucleation. Similar equations for progressive nucleation and other geometries appear in Table 3.1. The last step in the calculation is to account for overlap. Avrami[4] shows how the calculation is effected. If the nuclei overlap randomly as in Fig. 3.1.

$$S_1 = 1 - \exp\left(- S_{1\text{ex}}\right) \tag{3.10}$$

TABLE 3.1. Current into isolated nuclei (without overlap)

t	$t^{1/2}$	$t^{3/2}$	t^0
(1) $i = \dfrac{2ZF\pi M}{\rho} N_0 h k^2 t$	(4) $i = \dfrac{2F\rho}{M} N_0 \pi \theta^3 D^{3/2} t^{1/2}$	(5) $i = \dfrac{ZF\rho}{M} (\pi\theta^3 D A^{3/2}) \tfrac{2}{3} t^{3/2}$	(6) $i = \dfrac{ZF\rho}{M} \pi\theta^2 D$
(a) Instantaneous	Instantaneous	Progressive	Instantaneous
(b) 2D	3D	3D	2D
(c) periphery	periphery	periphery	periphery
(d) fast	slow	slow	slow
(2) $i = ZFAL^2 kt$			
(a) progressive			
(b) 1D needle			
(c) end (cross section L^2)			
(d) fast			
(3) $i = \dfrac{ZFh\rho}{M} \pi\theta^2 D A t$			
(a) progressive			
(b) 2D			
(c) periphery			
(d) slow			

TABLE 3.1 (cont.)

t^3	t^2	$\exp(t)$
(7) $i = \dfrac{2ZF\pi M^2}{3\rho^2} Ak^3 t^3$	(8) $i = \dfrac{2ZF\pi M^2}{\rho^2} N_0 k^3 t^2$	(10) $i = \dfrac{ZFk4LM}{\rho}\left[\exp\left(\dfrac{4kM}{L\rho}(t-t_0)+\ln r_0\right)\right]$
(a) progressive	instantaneous	instantaneous
(b) 3D	3D	1D (needle cross section L)
(c) periphery	periphery	round long axis ($r_0 ht$ at t_0).
(d) fast	fast	fast
	(9) $i = \dfrac{ZF\pi Mh}{\rho} Ak^2 t^2$	(11) $i = \dfrac{ZFkM}{\rho} 2\pi x_0\left[\exp\left(\dfrac{kM}{\rho}x_0(t-t_0)+\ln r_0\right)\right]$
	(a) progressive	instantaneous
	(b) 2D	3D
	(c) periphery	base periphery into section x_0 high (r_0 is radius of hemisphere at t_0, $r_0 > x_0$).
	(d) fast	fast

(a) nucleation type (b) growth type (c) site of slow step (d) diffusion

$S_{1ex} = \pi r^2$ is the top area of an isolated nucleus, and S_1 is the actual area when overlap is taken into account. Equation (3.9) becomes, for circular two-dimensional nuclei.

$$V = S_1 h = h\left(1 - \exp\left(-S_{1ex}\right)\right) \tag{3.11}$$

$[(dV/dr)]$ is known from equation (10) and $dr/dt = 2\pi r$ hence the current can be calculated, equation (3.2), for instantaneous or progressive nucleation. The results are

$$i = \frac{nF\,\pi M}{\rho}\,hA\,k^2\,t^2\,\exp\left(-\frac{\pi M^2\,A\,k^2\,t^3}{3\rho^2}\right) \tag{3.12}$$

for progressive and

$$i = \frac{nF\,\pi M}{\rho}\,N_0\,k^2\,t\,\exp\left(-\frac{\pi M^2\,N_0\,k^2\,t^2}{\rho^2}\right) \tag{3.13}$$

for instantaneous nucleation.

Equations (3.12), (3.13) have the characteristic shape shown in Fig. 3.2. The maximum current (i_m) and time (t_m) are given directly by differentiating

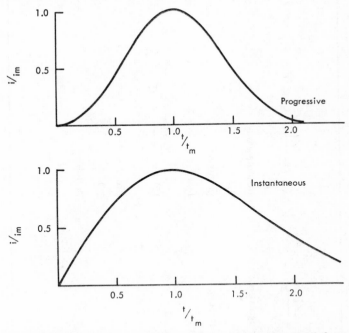

FIG. 3.2. Theoretical current line curves according to: (a) equation (3.12) for progressive nucleation; (b) equation (3.13) for instantaneous nucleation.

(3.12) or (3.13). Theoretical value of $i_m t_m / Q_m$ is independent of the potential dependence of k, and is 1·0 for progressive and 0·6 for instantaneous nucleation. The constancy of $i_m t_m / Q_m$ with potential is a useful experimental test.

The arguments used to derive equations (3.12), (3.13) can be extended to layer by layer growth[5] shown in Fig. (3.3). Fig. (3.4) displays the potentiostatic current–time transient.

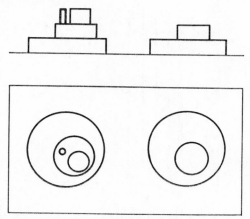

FIG. 3.3. Formation of a three-dimensional deposit by successive monolayer growth.[5]

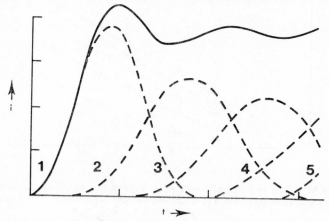

FIG. 3.4. Total current transient for layer by layer growth in a potentiostatic experiment, showing contribution from individual layers.[5]

(ii) *With diffusion*

The model of Fig. 3.1 will be considered. Assume that the number of nuclei formed initially, N_0, is large[6] and the diffusion zones around each nucleus have overlapped very soon after the potentiostatic pulse is started. It can then be assumed, as a limiting situation, that planar diffusion operates from $t = 0$ parallel to the electrode surface. The surface becomes progressively covered with deposit until a monolayer is formed and the current stops. The predicted current–time transient is similar to the adsorption of organics[7] and is shown in Fig. (3.5).

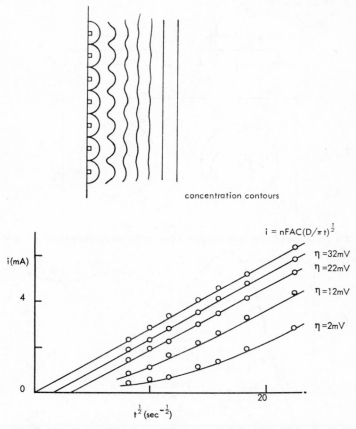

concentration contours

$$i = nFAC(D/\pi t)^{\frac{1}{2}}$$

FIG. 3.5. (a) Model for overlapped nuclei, growth controlled by planar diffusion. The full lines are concentration contours; (b) calculated current–time transient in a potentiostatic experiment.[6]

Another type of limiting situation can be calculated if the growth rate for a single nucleus growing by diffusion is subjected to the Avrami[4] treatment. It has been shown that a single nucleus under these conditions[8] grows as

$$r = \theta(Dt)^{\frac{1}{2}} \tag{3.14}$$

where θ is a constant controlled by the potential, M, ρ. The resulting relations, after applying equations (3.1), (3.2), (3.6), (3.10), are[5]

$$i = q_{mon}\, \pi\theta^2\, D \exp\left(-\frac{\pi\theta^2\, DAt^2}{2}\right) \tag{3.15}$$

for progressive nucleation and

$$i = q_{mon}\, \pi\theta^2\, D \exp\left(-\pi\theta^2\, DN_0\, t\right) \tag{3.16}$$

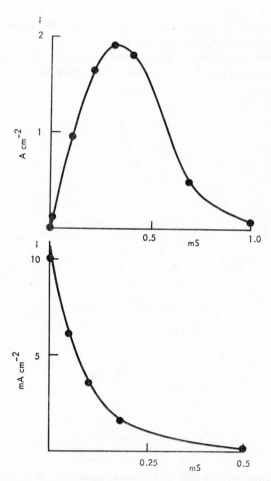

FIG. 3.6. Growth of two-dimensional layers controlled by diffusion (a) progressive nucleation according to equation (3.14); (b) instantaneous nucleation according to equation (3.15).[5]

for instanteneous nucleation. The form of equations (3.15), (3.16) is shown in Fig. (3.6).

The current–time relation can also be estimated when the nuclei are surrounded by a fixed diffusion zone[9]. A graph-paper simulation shows that $(i_m t_m / Q_m)$ is no longer a fixed constant but is less than the values for progressive nucleation, and for instaneous nucleation. The chemistry and electrochemistry of the system will determine how the apparent fixed diffusion zone arises.

(iii) *Other models*

It is appropriate to mention here other models, for three-dimensional growth. In principle three-dimensional deposits can always be formed by layer-by-layer growth.[5] However it is useful to consider a three-dimensional model[10] which ignores the finite size of atoms, when the spreading of two-

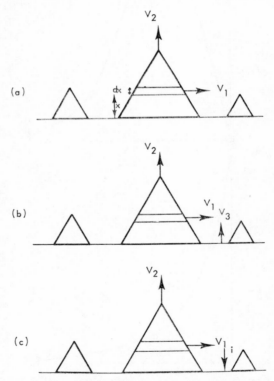

FIG. 3.7. Three-dimensional models (a) growth of circular cones on an inert substrate (metal deposition); (b) growth of circular cones on a substrate of the same material. The base plane moves with velocity V_3 (metal deposition); (c) current only flows into the uncovered area (passivation).[10]

dimensional layers is inhibited on a heterogeneous substrate for example. Fig. 3.7 shows a number of such models; the corresponding current–time curves are shown diagrammatically in Fig. 3.8. The calculation is performed by considering a slice dx and applying the equation (3.12), (3.13) for two-dimensional growth and overlap in the plane x. Integration over x gives the curve of Fig. 3.8.

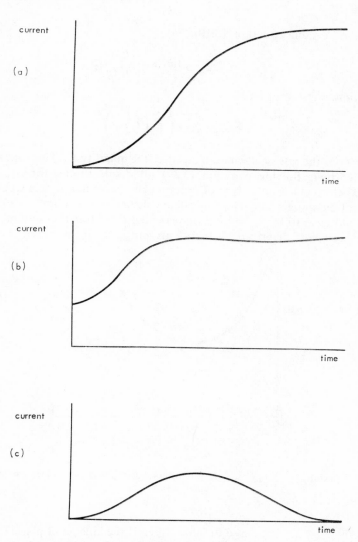

FIG. 3.8. The form of current–time transients for the models of Fig. 7.

CURRENT STEP METHOD

Only one situation has been calculated for this case[5]. Consider the model of Fig. 3.1. If a constant number of nuclei N_0 are formed at $t = 0$ then constant current i flowing generates

$$S_1 = \frac{itM}{nFl\rho}.$$

(3.17)

From the Avrami equation (3.10),

$$S_{1ex} = -\ln\left(1 - \frac{itM}{nFl\rho}\right).$$

(3.18)

In addition

$$S_{1ex} = N_0 \pi\left(\int_0^t V(t)\,dt\right)^2$$

(3.19)

where $V(t)$ is the rate of advance of an edge; the quantity $V(t)$ is proportional to the current. Solving equation (3.19) after substituting for S_{1ex} gives Fig. 3.9, which is a familar type of experimental curve not always recognised as nucleation and growth. The equivalent calculation for progressive nucleation would have to be carried out numerically. It is clear that potentiostatic pulse experiments described on p. 115 are far easier to interpret.

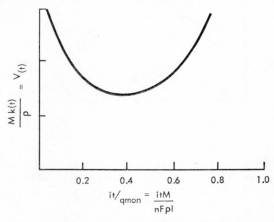

FIG. 3.9. Dependence of $V_{(t)}$ on time for instantaneous nucleation under galvanostatic conditions.[5]

A.C. METHOD

The method has not been used to follow directly the kinetics of phase forming reactions. However it is invaluable in detecting the presence of metallic or non-metallic thin films in conjunction with the sweep method. Three experimental situations have been distinguished:

(i) when a thin anodic film has measured capacity different from the underlying metal. The range of stability of the monolayer can be determined and the point at which multilayer growth starts.[11] It is also possible to investigate the properties of the film as a function of potential.

(ii) a monolayer metal film can be detected either if it has intrinsically a different double layer capacity than the underlying metal or if its presence impedes a dissolution reaction of the substrate metal.[12]

(iii) the fact that dissolution reactions leading to solution soluble complexes have a typical frequency response to an a.c. signal (see Chapter 2, p. 73) means that these reactions can be identified in the presence of a film.[13]

Table 3.2. gives examples of each type.

NUCLEATION AND GROWTH MODEL (WITH EDGE EFFECTS)

POTENTIAL STEP METHOD

(i) Without diffusion

Two models have been described. In the first one a single nucleus is initiated in various positions around the periphery of the electrode, Fig. 3.10. The current at any time is proportional to the growing area as before. In this case the authors define[14]

$$i = k\eta\, L(r) \tag{3.20}$$

and by simple geometrical arguments predict current time transients shown in Fig. 3.11; the transients of Fig. 3.11 are similar to those observed in practice.

The second model is more general;[15] it considers only that the growth of a single nucleus can be limited. The reason for nuclei reaching a maximum

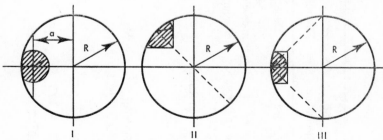

FIG. 3.10. Model to describe single nucleus formation at the electrode periphery.[14]

TABLE 3.2

Electrolyte	System Electrode	Deposit	Mechanism	Parallel dissolution as	Reference
Purified Ag NO$_3$	"Dislocation-free" ideal Ag microelectrodes	Ag	Successive monolayers nucleated from the edge.		14, 23
Ni^{2+} in KSCN	Mercury	Monolayer of Ni or Ni/Hg alloy	Single monolayer.		9
Co^{2+} in KSCN	Mercury	Monolayer of Co/Hg alloy + Co amalgam + multilayer of Co metal	At low η a monolayer, at higher η a multilayer.		24
Pt$^{\prime}$(iv) or Ru(ii) in HCl	Mercury	Monolayer of Pt/Hg alloy, Ru/Hg alloy	Monolayer formation accompanied by H$_2$ evolution.		12
Pb^{2+} + KCl	Ag	Pb	Monolayer and multilayer.		6
Tl$^+$ + KCl	Ag	Tl	Monolayer and multilayer.		6
cd^{2+} NaClO$_4$	Pb	Cd	Multilayer.		6

TABLE 3.2 (cont.)

Electrolyte	System Electrode	Deposit	Mechanism	Parallel dissolution as	Reference
KCl, HCl	Mercury	Hg_2Cl_2	At low η one multilayer, at high η a succession of monolayers are formed.	$HgCl_4^{2-}$	25
NaOH	Mercury	HgO	One monolayer + one multilayer.	$Hg(OH)_2$	26
HPO_4^{2-}	Mercury	Hg_2HPO_4	One monolayer + one multilayer.		27
Oxalate	Mercury	Hg oxalate	One Monolayer + one multilayer.		28
Barbituric acid	Mercury	Hg barbiturate	One monolayer.		29
S^-	Mercury	HgS	Two monolayers.	HgS_2^{2-}	30
KSCN	Mercury			$Hg(SCN)_2^-$$_2$	31
KCl	Tl(1% amalgam)	Tl Cl	Two monolayers and one multilayer.	Tl^+	32
NaOH	Cd (1% amalgam)	$Cd(OH)_2$	Three monolayers.	$Cd(OH)_4^{2-}$	33
NaOH	Solid Ag	Ag_2O	One monolayer and one Multilayer.	$Ag(OH)_2^-$	34, 35 (a, b)

size could be interaction with the boundary, growth on a limited site, or the adsorption of organics at the edge so that growth is stunted. If the time for the first nucleus to reach maximum size is t' then the nuclei have grown normally under random progressive nucleation up to that point and equation (3.12) is valid.

$$i_{t \ll t'} = \frac{nF \, \pi Mhk^2 \, At^2}{\rho} \exp \left(- \frac{\pi M^2 \, k^2 \, At^3}{3\rho^2} \right). \tag{3.21}$$

For $t \gg t'$ it can be shown that

$$i_{t > t'} = \frac{nF \, \pi Mhk^2 \, At'^2}{\rho} \exp \left[\frac{\pi M^2 \, k^2 \, A}{\rho^2} \left(\tfrac{2}{3} t'^2 - t'^2 t \right) \right]. \tag{3.22}$$

Equations (3.21), (3.22) are plotted in Fig. 3.12, where they are compared with a graph paper simulation in which the nuclei are allowed to interact with the boundary. It is clear that if $t' < t_m$, equations (3.21), (3.22) reproduce a situation in which nuclei interact with the edges of the electrode and with one another.

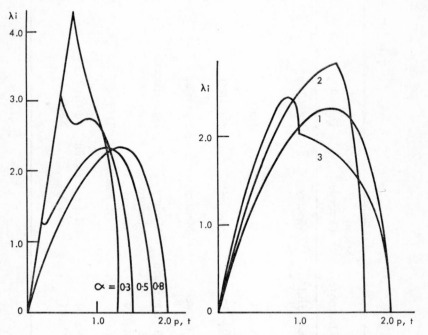

FIG. 3.11. Current–time curves for the potential step applied to the models of Fig. 10.[14]

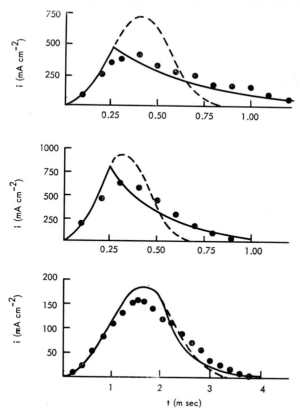

FIG. 3.12. Equations (3.21), (3.22), the full line compared to a graph paper simulation, denoted by the points ●.[15] The dashed line is calculated from equation (3.12).

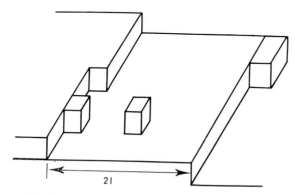

FIG. 3.13. Features present on a real metal surface.

ADATOM MODEL

The model is of theoretical interest and has not been observed in practice, although it has commonly been assumed to occur.

Based on vapour phase deposition of metals and surface studies it is fairly certain that the features of Fig. 3.13 are present on a metal surface. The step lines are probably part of a screw dislocation which grows as a self perpetuating structure without the need for nucleation. However the exact nature of the step lines does not need to be further specified. The essence of the adatom model is shown in Fig. 3.14. The electrochemical reaction produces adatoms (the adsorbed atoms of Fig. 3.13) which diffuse along the surface to the step lines where they are incorporated into the lattice. It is assumed that the step line edge does not advance significantly during the measurement. The mathematics is straight forward[16] but tedious so only the result will be given. Fig. 3.15 shows the result of applying a potentiostatic pulse (see ref. 16 for the complete theory) to the adatom model of Fig. 3.14 the curves are

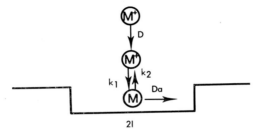

FIG. 3.14. The essentials of the adatom model.

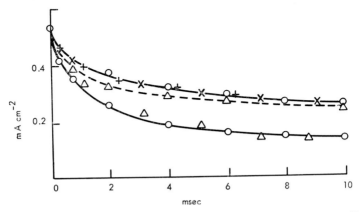

FIG. 3.15. The effect of a potentiostatic pulse on the model of Fig. 3.14.[17]

presented[17] as a function of the parameter $k^2 l/D$. The upper curve corresponds to capture by the step line before significant back reaction has occurred and the lower curve corresponds to back reaction before surface diffusion. The theory for the model of Fig. 3.14 for two other types of experiment the a.c.[18] and the galvanostatic method[19] have also appeared in the literature; the galvanostatic case can only be calculated in the absence of diffusion in the solution. The microstructure of Fig. 3.14 is inevitably superimposed on a macrostructure shown in Fig. 3.16. The effect of the macrostructure on the measurement of transients has been very fully considered by de Levie[20] and the reader is referred to his review.

FIG. 3.16. Model of a rough metal surface.[20]

A final comment on the adatom model is that the situation of Fig. (3.14) is unlikely to be stable. As deposition proceeds and the adatom concentration builds up on the surface it would seem possible that material will be directed into the step line. The ultimate situation can be envisaged in Fig. (3.17) where the step lines have become surrounded by diffusion zones; analysis of this model[21] has appeared in the literature for the steady state. However the theory of transition from the models of Figs. 3.14 to 3.17 has not yet been considered.

FIG. 3.17. Sink model of a surface.[21]

FORMATION OF THICK FILMS

In the case of oxide films[22] the current can be controlled by the diffusion or migration of either metal or oxide ions through the film. This situation arises when the thick films are grown. The study of the growth characteristics depends on the solid state properties of the films and outside the scope of this book. For further information the reader is referred to a review of the subject.[22]

EXPERIMENTAL RESULTS

The systems which have been investigated using the ideas of this Chapter are reviewed in the Table 3.2. The systems are metal deposition and battery reactions and are designed to investigate some feature of phase formation. The deposit was mainly identified by electron diffraction and the techniques of Chapter 4.

A potentiostatic pulse measurement generally shows the sequence of events given in the column headed mechanism. In some cases the monolayer equilibrium potential occurs before the multilayer value, and the monolayer is stable in a limited potential range without further growth. The a.c. method has been used extensively to confirm the stability of monolayers without further growth.

REFERENCES

1. Erdey–Gruz, T. and Volmer, M. *Z. Physik. Chem.* **157,** 165 (1931).
2. Harrison, J. A. and Thirsk, H. R. "Electroanalytical Chemistry", Vol. V (Ed., A. J. Bard), Marcel Dekker, New York (1971).
3. Fleischmann, M. and Thirsk, H. R. "Advances in Electrochemistry and Electrochemical Engineering", Vol. 3 (Ed., P. Delahay), John Wiley, New York (1963).
4. Avrami, M. *J. chem. Phys.* **7,** 1103 (1939); **8,** 212 (1940); **9,** 177 (1941).
5. Armstrong, R. D. and Harrison, J. A. *J. electrochem. Soc.* **116,** 328 (1969).
6. Astley, D. J., Harrison, J. A. and Thirsk, H. R. *J. electroanal. Chem.* **19,** 325 (1968).
7. Reinmuth, W. H. *J. phys. Chem.* **65,** 473 (1961).
8. Frank, F. C. *Proc. Roy. Soc.* **A201,** 586 (1950).
9. Fleischmann, M., Harrison, J. A. and Thirsk, H. R. *Trans. Far. Soc.* **61,** 2742 (1965).
10. Armstrong, R. D., Fleischmann, M. and Thirsk, H. R. *J. electroanal. Chem.* **11,** 208 (1966).
11. Armstrong, R. D. and Milewski, J. D. *J. electroanal. Chem.* **21,** 547 (1969).
12. Giles, R. D., Harrison, J. A. and Thirsk, H. R. *J. electroanal. Chem.* **20,** 47 (1969).
13. Armstrong, R. D., Race, W. P. and Thirsk, H. R. *J. electroanal. Chem.* **14,** 143 (1967).
14. Budewski, E., Bostanoff, W., Witanoff, T., Stoinoff, Z., Kotzewa, A. and Kaischew, R. *Electrochim. Acta.* **11,** 1697 (1966).
15(a). Oldfield, J. W. Ph.D. Thesis, University of Newcastle-upon-Tyne (1967).
15(b). Armstrong, R. D., private communication.
16. Fleischmann, M., Rangarajan, S. K. and Thirsk, H. R. *Trans. Far. Soc.* **63,** 1240, 1256 (1967).
17. Harrison, J. A. *J. electroanal. Chem.* **18,** 377 (1968).
18. Fleischmann, M., Rangarajan, S. K. and Thirsk, H. R. *Trans. Far. Soc.* **63,** 1251 (1967).
19. Rangarajan, S. K. *J. electroanal. Chem.* **16,** 485 (1968).

20. de Levie, R. *In* "Advances in Electrochemistry and Electrochemical Engineering", Vol 6 (Eds., P. Delahay and C. W. Tobias), Interscience, New York (1967).

21. Fleischmann, M. and Harrison, J. A. *Electrochim. Acta* **11,** 749 (1966).

22. Young, L. "Anodic Oxide Films", Academic Press, London and New York (1961).

23. Witanoff, T., Sevastyanoff, E., Stoinoff, Z. and Budewskii, E. *Sov. Electrochem.* **5,** 218, 411 (1969).

24. Astley, D. J. and Harrison, J. A. *Electrochim. Acta* **15,** 2007 (1970).

25. Armstrong, R. D., Fleischmann, M. and Thirsk, H. R. *Trans. Far. Soc.* **61,** 2238 (1965).

26. Armstrong, R. D., Race, W. P. and Thirsk, H. R. *J. electroanal. Chem.* **19,** 233 (1968).

27. Armstrong, R. D., Fleischmann, M. and Oldfield, J. W. *J. electroanal. Chem.* **14,** 235 (1967).

28. Armstrong, R. D. and Fleischmann, M. *Z. phys. Chem.* **52,** 131 (1967).

29. Armstrong, R. D. and Oldfield, J. W., private communication.

30. Armstrong, R. D., Porter, D. F. and Thirsk, H. R. *J. phys. Chem.* **72,** 2300 (1968).

31. Armstrong, R. D., Harrison, J. A. and Thirsk, H. R. *Corrosion Sci.* **10,** 629 (1970).

32. Fleischmann, M., Pattinson, J. and Thirsk, H. R. *Trans. Far. Soc.* **61,** 1256 (1965).

33. Armstrong, R. D., Milewski, J. D., Race, W. P. and Thirsk, H. R. *J. electroanal. Chem.* **21,** 517 (1969).

34. Giles, R. D., Harrison, J. A. and Thirsk, H. R. *J. electroanal. Chem.* **22,** 375 (1969).

35(a). Giles, R. D. and Harrison, J. A. *J. electroanal. Chem.* **24,** 399 (1970).

35(b). Giles, R. D. and Harrison, J. A. *J. electroanal. Chem.* **27,** 161 (1970).

Chapter 4

Ancillary Techniques

INTRODUCTION

The preceding chapters have been concerned exclusively with electrochemical parameters and their interpretation. The majority of electrochemical investigations, however, require consideration in three additional areas. They are (a) cell design and solution preparation; (b) analysis of the bulk solution; (c) the structure and structural changes of the electrode itself and experimental investigations of the electrode–solution interphase.

We do not intend to treat (b); skills in this experimental area are of particular importance in mechanistic studies and exercises in electrochemical synthesis. This point is generally well understood and, in the best work, adequate use is made of, often necessarily sophisticated, techniques of chemical analysis; the methods involved are not peculiar to electrochemistry, excepting in that methods of sampling will affect cell design, and they lie within the general chemical experience.

Our main concern is with (c) and the outlining of experimental approaches to studies in this area which often are extremely valuable in giving quite independent support to proposals developed from the analysis of the electrical parameters. In the past few years there has been a very welcome trend for developments of interest in this direction, which in the past has been neglected, possibly in part because the necessary instrumentation and knowledge is less immediately accessible to the majority of electrochemists than, say, skills in analysis, and in part because instrumental development has, itself, been very rapid.

CELLS

The sophistication or otherwise of the cell used in an electrochemical investigation is so much a feature of the experiment involved that the design is always a matter for individual consideration. Factors involved require careful *a priori* assessment and include some, but fortunately not all of the following desiderata; facilities for the purification of the solvent and possible extraction at the termination of an experimental run; admission of gases with attendant

136

purification trains; exclusion of gases, particularly oxygen; sampling arrangements for products; precise setting of the working electrode and, with a three-electrode cell, the associated Luggin capillary; elimination as far as possible of liquid junctions leading to unknown contributions either to a measured or a controlled overpotential; separation of working electrode compartment from ancillary electrode compartment without the introduction of an excessive resistance into the system; arrangements for optical or X-ray examination of the electrode *in situ* for the examination of the electrode surface; arrangements for optical devices for monitoring shifts in solution concentrations, diffusion layer, adsorption; precise control of mercury electrode, i.e. dropping mercury electrodes, hanging drop of supported drop systems; arrangement for monitoring solid electrodes in a convenient way for subsequent removal for structural examination; devices for removing films from liquid, mercury or mercury amalgam substrates; devices for breaking sealed electrodes *in situ*, prior to an experimental run.

The arrangement of the electrodes in a cell for use with fast response time instrumentation is important and has been discussed at length;[1] the problem of design of cells for use in a.c. bridges depends on the frequency range employed.[2] Temperature control can also be troublesome with some experimental arrangements but a well-designed air thermostat is a most useful device for many cells.

Solvent and solvent preparation is, as in any physical chemistry experimentation, a matter requiring care; for organic solvents for example there are several comprehensive references;[3(a,b)] good conductivity water should be employed and rigid procedures in order to purify all chemicals used; some authors advise pre-electrolysis for final purification of electrolytes prior to experimental runs.

Most metal electrodes require careful cutting polishing and cleaning; single crystals are best cut by spark techniques[4] or by a chemical saw[5(a,b)] followed by chemical or electrochemical polishing or etching;[5(c)] there will be some further discussion of these points later in the chapter. There is a helpful summary of many experimental points in an article by Bockris and Damjanovic.[6]

It is important to maintain a sense of proportion in experiments containing many factors of uncertainty and time should not be wasted on isolated aspects of technique with the arbitrary exclusion of other factors.

OPTICAL PROCEDURES AND THE EXAMINATION OF SURFACES DURING
KINETIC STUDIES

Optical methods have the great advantage in that they lend themselves particularly well to *in situ* studies during the course of the kinetic investigation. They may be grouped as follows: microscopy for the study of surface topography; interference methods for thickness determination in the study of growing uniform layers; Schlieren techniques depending on changes of refraction index of zones of changing concentration adjacent to the electrode face; reflection of polarised light, ellipsometry and reflection spectroscopy.

Interferences and Schlieren techniques do not seem to have been productive of much electrochemical interest now for a number of years and will not be further discussed. Information on experiment and theory may be found in a few general references.[7] It would seem that the method of growth of thin and optically transparent films limits the application of interferometric measurements: schlieren methods concern the diffusion layer alone; this in, for example, the case of high current density processes such as electroforming might be of interest. Fundamentally, elucidation of electrode phenomena is relevant to a more micro situation.

FIG. 4.1. Photomicrograph of the cross section of a nickel hydroxide layer 3×10^{-3} cm thick and detached from the supporting electrode after partial charging. Shows separate diffuse and sharp oxidation boundaries advancing from the metal (lower) to solution (upper) side of the layer.

Optical microscopy

Optical microscopy has been used intermittently for the study, in particular, of deposition processes and the technique has been successfully developed through the creation of cells such as that described in considerable detail in the reference cited above[6] where careful design can maintain adequate electrolyte flow with a correct objective-coverslip-specimen geometry.

Examples of conventional microscopy are readily found but it still can be surprising to find the information that can be revealed by careful microscopy.

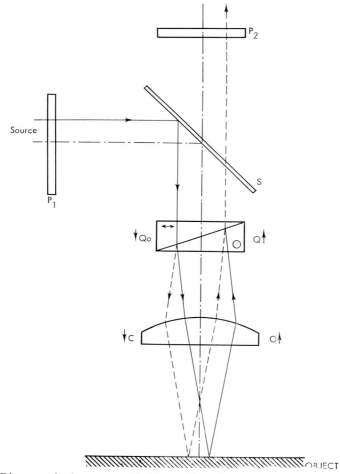

FIG. 4.2. Diagram of microscope with Nomarski optics for reflected light. P_1, P_2— Polariser and analyser; S—half-silvered mirror; Q_0, O—Modified prism acting as beam splitter (orthogonal beams) and compensator. Function dependent on wave direction. C, O—Condenser and objective.

For example, the process of reduction of nickel hydroxide can be observed with surprising clarity; the authors are indebted to a colleague for Fig. 4.1, illustrating this point.[8] The skill here lies rather in the careful organization of the material than in the sophistication of the microscopy.

The possibilities of using a long working distance, with special objectives, flexible means of illumination and optical interference techniques greatly extend the use of optical microscopy. The differential phase contrast technique, that is, the employment of Nomarski[9] optics, is of particular interest for electrode studies; this method has been subjected to considerable analysis and instrumental development over the last fifteen years; appropriate optics can now be added to many research microscopes.

The method is based on the splitting of plane polarised light by a modified Wollaston prism, the auxiliary prism, into two components which travel parallel, but closely spaced, to each other either through or by reflection to and from a reflecting specimen; the direction along which the beam is sheared is an important reference in examining surfaces. A second Wollaston prism, the main prism, actually the same one with the metallurgical microscope and located in the focal plane of the objective, recombines the two components which are polarised at 90° to each other. A final analyser makes the vibration planes parallel so that the bundles of light may interfere. The microscope image appears to gain a third dimension from the generation of a gradient of phase difference across the object in the direction of shear, the optical object is perceived by its slopes; the geometrical configuration is thus more important than the absolute phase shift it induces as a generator of contrast. It is thus also necessary to be able to rotate the specimen surface in its own plane. The technique is one for which a very sound knowledge of the principles involved is necessary both for satisfactory results and correct interpretation. Several references[9] of a fundamental type are included to give some insight into the use of the technique in transmission and reflectance. Fig. 4.2 is a schematic diagram for the optical system suitable for application to a horizontal electrode system. Bockris[10] has applied this method to the study of copper electrodeposition and Fig. 4.3 is taken from his work illustrating a copper electrodeposited surface as seen by optical microscopy with vertical illumination and by the differential phase contrast technique. The smoothness of the surface is much more readily observed by the latter method but it should be remembered that the reference plane is formed optically and can be tilted arbitrarily to the specimen surface, therefore deduction as to the smoothness originates from the waviness of the fringes rather than their spacing. The method is sensitive and can detect steps with an optical difference of as little as $\lambda/10\phi$; a compensator of known angle enabling an accurate assesment of this difference in height to be made.

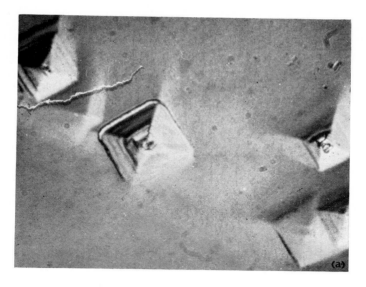

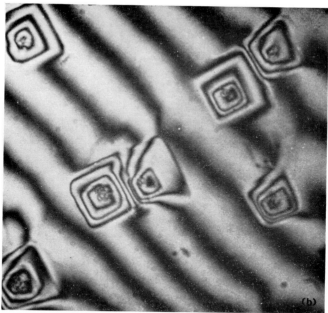

Fig. 4.3. Electrodeposited copper seen by (a) vertical illumination (bright field),[10] (b) differential phase contrast.

Reflection of polarised light, reflection spectroscopy and ellipsometry

The situation, to-date, in the application of these optical methods with rather extensive reference to electrode processes has been summarised by the proceedings of a Faraday Society Symposium.[11] Interest in exploiting these extremely sensitive methods, for information is obtainable even from incomplete mololayers, has developed considerably in the last few years. All the methods differ from optical or interferometric techniques in that they depend more substantially on energy absorption from the light source and sensitivity is thus only limited by the identification of a charge in the measured signal associated with inescapable "noise" or random signals associated with the experiment.

The best known and oldest method is that of ellipsometry, and this will be dealt with first, although the reflection of polarised light and reflection spectroscopy are in many ways more direct in the possibilities of interpretation.

Ellipsometry

This is a non-destructive method, which is useful for both the study of extremely thin films and for accurate measurements of refractive index and thickness of thicker films. It is however a difficult experimental technique and certainly requires access to computer facilities for handling the data since the associated calculations demand extended treatment; development for certain kinetic observations would seem to be feasible and certainly desirable. The electrode surface is readily examined *in situ*.

The method depends on the evaluation of the change in the state of polarisation of light, originally elliptically polarised, when reflected from a surface prior to, and subsequent to film formation; there is a most extensive literature although applications, to-date, to electrochemical studies are not yet extensive and mainly refer to passivity studies. What still seems to be required is a correlation of ellipsometric results, and indeed reflectance measurements in

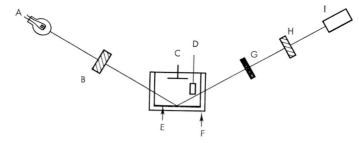

FIG. 4.4. Diagram of optical scheme for ellipsometric studies on electrodes.[17] (A) Monochromatic stabilised light source; (B) polariser; (C) counter electrode; (D) reference electrode; (E) working electrode; (F) quartz cell; (G) $\lambda/4$ wave plate; (H) analyser; (I) photomultiplier.

general with specific information from other sources and, most probably, electron optical observations on *well characterised* films. In principle, it would seem that the technique is of value for studies of film growth, associated with changes in film properties, and for adsorption and double layer studies. It could well be that partial exploitation of the method would be helpful and, at this time, further instrumental development for automatic recording are essential for kinetic studies.

The basic theory was developed by Drude[12] in the period 1889/90 for the evaluation of optical constants. The procedure is to sum the amplitudes of the reflected beams from the two interfaces for the perpendicular and parallel component giving the amplitudes of the total reflection R_s and R_p.

The basic equation is given by

$$\frac{R_s}{R_p} = \tan \Psi e^{i\Delta} \qquad i = \sqrt{-1}$$

where $\tan \Psi$ is the amplitude ratio and Δ equals the phase difference between the components of the electric vector in the plane of incidence and normal to it. These factors are determined by the instrumental settings and have been dealt with at length by, for example. Winterbottom[13] and McCrackin *et al*[14] The interpretation of these experimental values lies in the fact that the right hand side of the equation is expressable in terms of the Fresnel reflection coefficients, functions of the refractive indices and the angles of reflection, the thickness of the film and the wave length of the light λ.

Earlier work made use of the simplified Drude equations relevant to conditions where the thickness of the film was much less than λ, the wave length of the incident light. With the use of computers the full expressions can now be used and graphically represented as Ψ/Δ curves for assumed values of the variables of interest. This may be plotted as a $\Delta \Psi$ or $\Delta/\tan \Psi$ presentation to see the concordance between experimental and computed data. Many aspects of the method, as applied to non-electrochemical slow film growth are very well developed, for example, most helpful accounts are to be found in the report of a fairly recent Symposium,[15] and a series of reports by Muller and collaborators;[16] reference to an important later symposium has been made above.[11]

Most observations for electrochemical purposes have been carried out on equipment with the general layout, sketched in Fig. 4.4 with the components mounted on two optical benches moving in a plane, vertical to the surface as used, for example, by Bockris and Reddy.[17] Necessary instrumentation for actual kinetic observations, *in situ*, would follow some scheme such as that outlined by Winterbottom,[15] Fig. 4.5.

Bockris and collaborators have recently applied ellipsometric techniques to a number of systems subjected to anodic polarisation. Two of interest, since they have been studied extensively by alternative methods, and thus the pitfalls, as well as the advantages of the method can be evaluated, are the calomel system[17] and the nickel–nickel oxide system in sulphuric acid;[18] as stated above, applications to studies of passivating layers are moderately frequent in recent literature. There is a great deal of further experience and development required before such studies become fully definitive although a very useful start has been made; the possible nature of these developments is as follows. Firstly, the deposits of material are often somewhat abnormal in composition and specific in orientation and do not necessarily correspond to carefully prepared bulk material on which optical and X-ray diffraction data are based. Indeed, there is good evidence, particularly with monolayers, from e.m.f. data that this is definitely the case. Therefore, although ellipsometric measurements can be made *in situ*, they must be associated with different techniques for confirmation of the nature of the deposited layers. In addition, the assumption of layers of uniform thickness is unrealistic; the theory for layers of varying thickness is developing, for example, work by Hayfield *et al.*,[19] but the difficulties in analysis appear considerable and other evidence must be sought to establish in each case the uniformity or otherwise of the deposit.

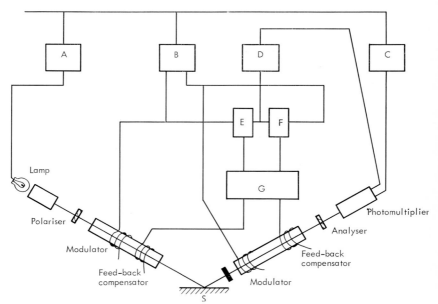

FIG. 4.5. Ellipsometric scheme for slow kinetic observations.[15] (A) Stabilised power source for lamp; (B) two phase generator; (C) stabilised high tension supply; (D) amplifier; (E, F) phase sensitive detector; (G) recorder; (S) electrode surface.

Another feature of the technique, which is still to be exploited results naturally from the fact that the vector notation depends on light absorption; interesting properties of the deposits may often result in changes in conductivity and similar properties which would be revealed by using light of differing wave lengths; it should be remembered that the majority of instruments are designed optically for a specific wavelength. Facilities for exploiting the possibility of using a range of wavelength could be incorporated into the instrumentation, even if the results are not immediately interpretable but merely indicative of non-classical behaviour. The improved sensitivity of equipment would seem to make it possible in principle to assume that all aspects of the double layer structure both in the metallic and solution phase contribute to the signal. Actual interpretation is still very problematical; the general nature of the problem has been examined very recently by Stedman[11] with reference to equilibrium conditions.

Reflectance methods

Measurements are made to determine changes in the intensity $\Delta R/R$ of the reflected light for perpendicular and parallel polarisations. The sensitivity can be increased by modulation techniques as described by McIntyre[20] and Bewick and Tuxford.[11]

The technique is more easily adapted for measurements at varying wavelengths and for kinetic studies. Barrett and Parsons[11] have described an alternative instrument, which may also be used as an ellipsometer and which they have used for studies of adsorption on platinum of oxygen, hydrogen, halides, methanol, formaldehyde and formic acid. A good deal of further development and application is still clearly necessary in this field because results can be different from coulometric and radioactive tracer results. Results for halide adsorption, for example, agree with the ellipsometric results of Yung-Chech Chim and Genshaw[21] which were at variance with the coulometry carried out by Gilman[22] and radioactive tracer determinations by Bolashova and Kagarinov.[23]

The interpretation of the effects on reflectance by the potential modulation of the surface is receiving increasing attention.[24] Cahan et al.[11] have studied reflectance at gold electrolyte interfaces by single and multiple reflection experiments with particular reference to the determination of the mechanism responsible for the electromodulation of the reflectivity. This was originally interpreted as due to a change in the optical constants of the double layer, a point contradicted by Hansen et al.,[25] who interpreted the change in terms of the variation in the electron concentration in the surface layers of the metal.

The correct interpretation of these effects is still to be resolved, nevertheless the whole area is one of great interest to electrochemists.

Plieth,[11] examining the adsorption of and reactivity of dyestuffs at electrode surfaces has used both multiple reflection and modulated single specular reflection techniques and has advanced arguments for the advantage of the latter because of the lack of interference from the electrolyte.

Reflection spectroscopy as a means of studing adsorbed layers in electro-chemical reactions is receiving an increasing amount of both theoretical and experimental attention. Hansen[26] has developed general approximate equations representing reflection spectra which are simple functions of the optical characteristics of the adsorbed species; the most intense spectrum arises with the use of parallel polarisation alone, since any perpendicular component present acts like stray light. The method has been used for studying the formation and properties of very thin films in surfaces and the products of heterogeneous chemical and electrochemical reactions by Koch,[27,28] McIntyre[29] and Yeager et al.[30]

Spectroscopic studies, by quite a number of authors, have now been published using attenuated total reflection optics with transparent electrodes of a semi-conducting nature such as GaP, SiC, ZnO; some of these substances are only transparent in the infra-red region. References can be made to papers by Kuwana[31] and collaborators and with potential modulation by Memming and Müllens.[11]

Transparent electrodes have been made with glass or quartz and a thin metal film,[32] a thin metal grid[33] also, as with the work on the electrochemistry of adsorbed dye layers of Memming and Müllens, with doped tin oxide[34] and germanium.[35]

A general treatment of internal reflection spectroscopy is to be found in a recent book by Marrick.[36] Application of rapid scan spectrophotometry to electrochemical kinetic studies, in seeking advantages from increased sensitivity, rejection of background and improvement in resolution could well follow with the further development of a Fourier Transform spectrometer.[37]

X-RAY METHODS FOR *IN SITU* STUDIES

For slow depositions and phase changes, considerable success has been reported by some workers in the application of X-ray diffraction methods during deposition. Briggs[38] developed a very satisfactory modification of an X-ray powder camera with an oscillating wire electrode mounted in the place of the usual powder specimen on a single crystal goniometer, with a 6 mm diameter cylindrical film holder, which proved to be of considerable value in studies then being carried out on the nickel oxide electrode system. A vertical transfer of one was all that was required and an oscillation of 2 Hz suffices to prevent drying out of the electrode. This is illustrated in Fig. 4.6 with the associated intensity profile for an anodically formed nickel hydroxide specimen.

Somewhat later, rather more sophisticated equipment, using a partially submerged rotating electrode for use on a horizontal circle X-ray spectrogoniometer was successfully developed and used by Burbank and Wales.[39]

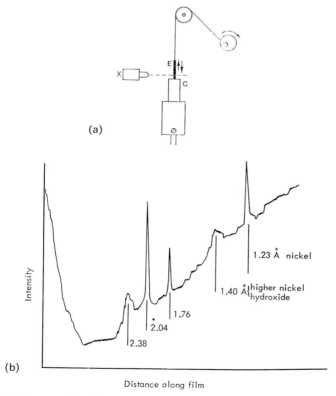

(a)

(b)

Distance along film

FIG. 4.6. (a) Diagram of oscillating electrode mounted on X-ray camera. (X) Source of X-rays; (C) cell; (E) electrode. (b) X-ray intensity profile taken during the anodic oxidation of nickel.

SURFACE EXAMINATION PRIOR AND SUBSEQUENT TO KINETIC EXPERIMENTS

An essential feature in the study of the kinetics of formation of surface deposits must be the characterisation of the material formed, both morphologically and as crystalline entities. The nature of the internal orientation, size of the crystals, the probable geometry of the growth, the existence of a preferred orientation with respect to the substrate, the effect of the substrate and the demonstration of the nature of the nucleation of the deposited phase are all matters which are highly relevant to the kinetics and mechanism. It is unfortunately true that this kind of information properly related to kinetic data is still very incomplete, even with certain well studied systems, and scanty or absent in many systems of importance. The examples of such studies have been given in Chapter 3 where the kinetic procedures have been based largely on experimental methods designed to control nucleation and electro-chemical kinetic growth parameters, and to observe electrocrystallisation during the limited time range over which free growth can be observed and where the lattice building step may be examined without it being obscured by the onset of mass transfer to the surface and overlapping of the crystals. As a consequence of this, such deposits are either in the form of very thin films, or isolated crystallites, adhering to the solid electrode substrate. In certain cases these deposits may be examined by X-ray diffraction, but predominantly the exiguity of the material is such as to demand recourse to electron optical procedures. There are many objections that can be raised against these techniques; it is impossible, for example, to examine the material at the moment of formation; heating by the electron beam under the conditions of a comparatively good vacuum can lead to recrystallisation or decomposition, but perhaps not so readily as inexperience with these methods could lead one to think. Particularly relevant to this objection is the knowledge that many oxide and hydroxide systems possess important properties related to hydroxyl ions incorporated during electrochemical formation. Such materials, which often give very diffuse diffraction rings can heat up in the electron beam with consequent loss of water and the formation of a more sharply diffused oxide pattern. Nevertheless, at the moment, electron optical methods are still the most powerful ones available for these structural investigations and are constantly becoming more versatile. A great deal more information of a systematic kind necessary for the advancement of the kinetic studies is still to be obtained from carefully designed experiments.

The instruments to which the most specific reference is made are conventional electron diffraction, electron microscopy with or without associated electron diffraction, the scanning electron microscope, electron probe microanalysis and the analytical electron microscope.

ELECTRON DIFFRACTION AND CONVENTIONAL ELECTRON MICROSCOPY

Both the experimental and interpretive techniques associated with these electron optical instruments have undergone a long period of development and for their effective application in electrochemistry a broad knowledge of the procedures is quite necessary. As a guide to the interpretive problems a reference to a major text, that of Hirsch et al.,[40] will serve as a most adequate introduction to the analysis of electron diffraction and conventional electron microscopy data most relevant to the type of specimens electrochemical investigations can yield. Material must be in a certain form for successful examination and this can affect the electrochemical design of the experiment; for this reason certain sections below are described in rather more experimental detail than in the remainder of the book.

In more detail we can classify our varying specimens of electrochemical interest as follows:

(1) powders;

(2) thin films formed by electrodeposition, which can be removed from the electrode and studied by transmission electron diffraction or electron microscopy, e.g. anodic films on Hg and amalgams and Al_2O_3 stripped form Al using $HgCl_2$;

(3) solid electrodes mechanically and electropolished—the surface finish and structure being examined before using the electrodes for electrodeposition;

(4) electrodeposits formed on solid polycrystalline and single crystal and metal electrodes, which cannot be stripped from the electrodes, and thus require the use of glancing incidence electron diffraction for structure and replica techniques for electron microscopy.

(i) Powders

For the preparation of powder specimens it is helpful to utilize (80 kHz) ultra-sonic dispersion with the fine particulate matter suspended in (a) H_2O, (b) H_2O + ethanol or (c) isobutyl-alcohol. A drop of suspension is placed on carbon-stabilized formvar sustrate on electron microscopy grids using fine glass capillary; petroleum ether is especially suitable for dispersion for electron diffraction studies.

Unshadowed specimens are examined in the electron microscope using selective aperture diffraction (10μm, 2μm apertures); specimens may be shadowed in order to determine the shape. In certain cases, it may be helpful to section the material by embedding in methacrylate and sectioning by a suitable ultramicrotome using a glass knife. Good dispersion is obtained when the embedding material is dissolved away in amyl alcohol and the specimen then shadowed.

(ii) Removable thin films

Thin films are prepared on a clean mercury or amalgam pool in special cells, under the conditions of the kinetic study. After formation of the film, the electrolyte is drained off, the deposit is thoroughly washed with distilled water, and rapidly dried by flooding with ethanol. The electrodeposited film is then covered with a thin layer of formvar, by an injection through a side arm and without access to air and this is backed with a relatively thick layer of collodion. At this stage the surface of the mercury or amalgam pool is lowered, thus leaving the electrodeposit adhering to the plastic support, and this circular disc is removed from the cell by using a small scalpel or razor blade. Frequently, a deposit of carbon is evaporated in a coating plant on to the electrodeposit. There are two reasons for doing this; it reduces the effect of the electron beam on the specimens and also acts as a supporting membrane.

Small square pieces (side approximately 2 mm) are cut out of the composite disc and placed centrally on 3 mm diameter copper electron microscope specimen grids. The plastic backing films are dissolved away by placing the grids with the specimens on a coarse steel or nickel mesh in small weighing bottles containing firstly amyl acetate and left for half an hour, and then chloroform, for a quarter of an hour. Thus, the final electron microscope specimen consists of the grid plus a thin carbon support film, plus the electrodeposit.

The use of a tilting goniometer stage in the electron microscope is particularly valuable in the evaluation of orientation and stacking faults in the films. Thus, as an example of a well studied electrochemical system, Fig. 4.7 is the diffraction pattern from a tilted film of $Cd(OH)_2$, formed anodically from cadmium amalgam to confirm the CdI_2 structure with normal lattice spacings in the directions perpendicular to the electrode surface, the hexagonal basal plane of the crystals lying parallel to the amalgam surface.

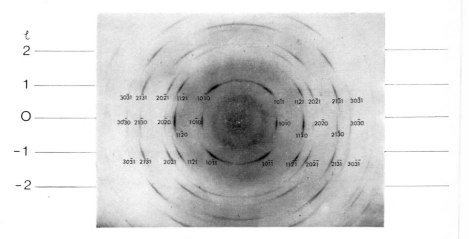

FIG. 4.7. Transmission electron diffraction pattern from an anodically formed cadmium hydroxide deposit. Partially indexed to identify the orientation (hexagonal basal plane parallel to the substrate) and the plane spacing vertical to the electrode. Angle of tilt with respect of the electron beam, 43°.

(iii) *Preparation of solid polycrysalline or single crystal metal electrodes*

The normal technique for preparing the surface of metal electrodes has been to cut a suitable size piece from a rod or block of material using a fine saw or chemical saw or by the methods referred to earlier in the chapter.[4] Cut surfaces are usually polished on emery papers, going from 0/0 to 6/0 grades with ethanol as a lubricant. The final polish may be obtained by using either fast or slow cutting alumina moistened with water on selvyt cloth placed on a flat glass plate, or diamantine paste on special wooden laps or felt pads. Electrodes are then electropolished or chemically polished[5c] before placing them in the electron diffraction equipment to ascertain their orientation. In the absence of a final careful electron diffraction study of the surface of single crystal faces it is impossible to guess at the surface organization produced by the preparative techniques. This certainly cannot be done by X-ray methods.

Fig. 4.8 shows an electron diffraction pattern taken at a very low angle of incidence from a carefully prepared face, close to a cube face, of a platinum single crystal; note particularly the absence of three-dimensional diffraction.

SURFACE REPLICATING METHODS

It is useful to remember that replicas often are useful for conventional microscopy as well as for electron microscopy.

(i) *Two-stage negative carbon replica*

The first impression is obtained from 1% or 2% collodion cast. This is

Fig. 4.8. Electron diffraction pattern of a single crystal platinum electrode surface. Very close to a cube face, cube edge azimuth. It will be noted that there is a complete absence of three-dimensional diffraction spots.

stripped on to specimen grid with Sellotape. The impression pre-shadowed usually at $\cot^{-1}2$ or $\cot^{-1}3$ with Au/Pd alloy (40/60) and then a thin layer of carbon evaporated on top of the shadowing metal. The collodion is dissolved away in amyl acetate.

(ii) *Unshadowed pseudo-negative replicas*

A 1% or 2% collodion stripping is taken from the electrodeposit. Carbon is evaporated on to the collodion replica and then the plastic dissolved in amyl acetate. Some crystals may be removed from the electrodeposit and these can be examined by transmission electron diffraction and electron microscopy prior to shadowing.

(iii) *Positive replicas*

The above process is carried out, but after the carbon is deposited and the plastic dissolved away the grids are *inverted* and the carbon film is shadowed through the grid bars, and thus a positive replica is obtained.

(iv) *Single stage positive replicas*

Single stage positive carbon replicas can sometimes be obtained from electrodeposits which are soluble in water by pre-shadowing the deposit and evaporating carbon on normally. On immersing the electrode in water, the replicating layer should float free and the shadowed carbon film be picked up on the grids.

(v) *and* (vi) *Other methods*

In other cases, perhaps where it is not permissible to use collodion or formvar to form a first impression, (due to chemical interaction as with, for example, MnO_2), a pre-shadowed carbon layer may be put down on to the deposit. The composite layer may then be removed from the electrode surface using sellotape.

A special two-stage negative replica technique[41] has proved to be of wide application. A primary impression of the surface to be studied is obtained by placing it in contact, under slight pressure, with thin 0·00, 10 μm, acetone softened cellulose acetate sheeting, Clarifoil. This impression is shadowed with Au/Pd and then coated with a thin layer of evaporated carbon. Small selected areas ($= 2 \times 2$ mm) are cut out of the sheeting and placed on electron microscope grids. The pre-shadowed carbon layer is strengthened with a drop of 0·1% Formvar and then Clarifoil is dissolved in acetone and the formvar removed with chloroform. This technique is especially suitable for awkward shaped specimens, such as thin wire, electrodes with a diameter of 2 mm or less, and for curved and thin cross sections of material, such as glass electrodes. For very rough specimens, in order to obtain a uniform layer of carbon over

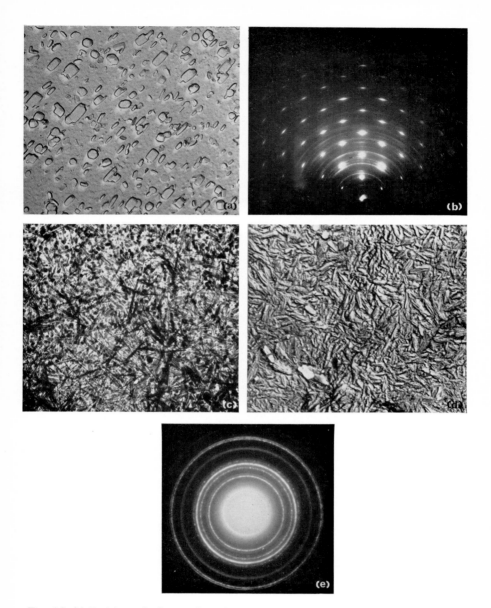

FIG. 4.9. (a) Positive collodion replica of an anodically formed Cd(OH)$_2$ deposit formed on the single crystal prism (10$\bar{1}$0) face of cadmium. Shadowed Au/Pd at cot$^-$,$^{1/3} \times$ 12,000. (b) Electron diffraction pattern from the same surface. Prism face (10$\bar{1}$0). [0001] azimuth. (c) An anodic deposit of Cu$_2$O and Cu(OH)$_2$ formed on copper "amalgam" and viewed by transmission electron microscopy, $\times$ 7,200. (d) Surface (carbon) replica of the same deposit $\times$ 7,200. (e) Electron diffraction pattern for identification. The shape of the diffraction spots will permit further analysis relating the internal crystalline structure of the crystals to the external shape.

the surface of "undercuts", it is necessary to rotate the Clarifoil impression during the carbon deposition which is carried out at an angle of, say 20° or 30° to the normal to the specimen, instead of at right angles as is usually the case. Some of these techniques are illustrated in Fig. 4.9 using the anodic oxidation of cadmium and cadmium and copper amalgam as examples.

SCANNING ELECTRON MICROSCOPY

Developments in the last few years of the scanning electron microscope has made available an extremely valuable instrument for association with electro-chemical studies. General reviews concerning the instrument are few but a

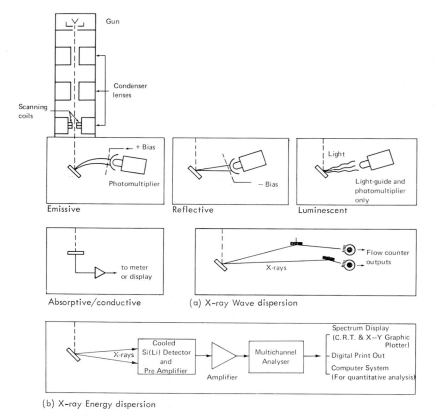

FIG. 4.10. Operating modes for a scanning electron microscope. The emissive and reflec-tive modes give rise to the micrograph image as does the transmission mode with the detector under the specimen which is not shown in this diagram. The X-ray emission analysis. (a) wave dispersion system, (b) energy dispersion (X-ray analysis).

book[42] is helpful, as is an extensive review;[43] a scanning electron microscope symposium is held annually and the proceedings published.[44]

The principle of the instrument is as follows. A primary beam of electrons from a heated tungsten filament is focussed into a fine probe and made to scan in a raster on the surface of interest, which may be the specimen as received or, if an insulator, i.e. hydroxide or salt, subjected to an *even* metal shadowing prior to positioning in the microscope; this requires a different procedure from the shadowing or a replica. The electrons liberated from the specimen by the probe are detected by a scintillator–photomultiplier system, the so-called emissive mode, or detected as a current under the specimen, that is, by transmission, the conduction mode; the resulting signals are used to modulate the brightness of a cathode-ray tube screen which is scanned in synchronism with the electron probe scanning the specimen. Secondary electrons are mainly used and details of deeply re-entrant holes are readily available. For similar magnifications the depth of focus is of the order of

FIG. 4.11. Scanning electron micrograph of Ag_2O formed anodically on a single crystal of silver, $\times$ 10,000.

three hundred times that of a light microscope, and resolution of 200 Å is readily realisable, indeed scanning electron microscopy is now in a position to compete with transmission microscopes for high resolution work; roughly, the working magnification range can be put at fourteen times to seventy thousand times. Other modes of operation are also available with additional instrumentation. Thus luminescence, detectable by measuring the optical emission, semi-conductor properties from absorption and conductivity and

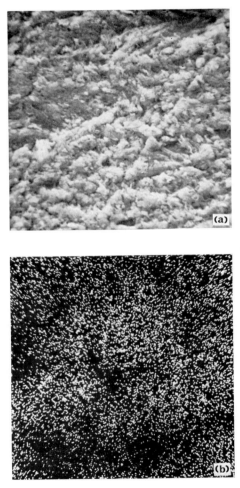

FIG. 4.12. Identification of a chromium compound, suggested from electrochemical kinetic observations, formed on a carbon electrode substrate. (a) electrode surface (scanning microscopy) ×600; (b) identification of chromium compound.

chemical identification from X-ray emission are all possible from appropriate specimens, and in addition selected area electron diffraction is also possible. These various operating modes are shown diagrammatically in Fig. 4.10, whilst Fig. 4.11, a typical micrograph, is from a specimen of anodically formed Ag_2O on silver which is also of ideal form for electron diffraction studies.

If the interest is in the semi-conducting qualities of electrochemically formed materials, it is worth noting that there has been recent progress with the production and measurement of an Auger electron spectra in a commercial scanning electron microscope with an improved vacuum system utilising ion pumping.[45]

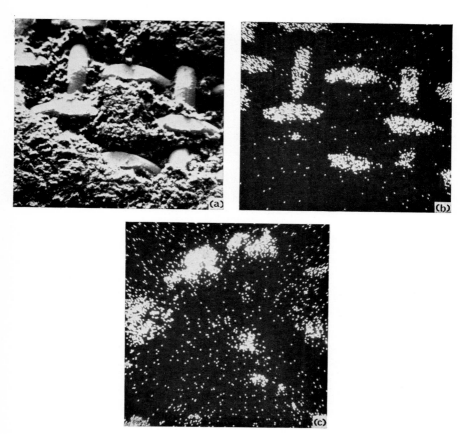

FIG. 4.13. Scanning electron micrograph of a grid-supported catalysed carbon electrode. (a) Micrograph × 60; (b) analyser set for nickel X-ray excitation; (c) analyser set for silver excitation.

The most important development in the use of the scanning electron microscope has been the development of energy dispersion X-ray analysis.[46(a,b)]

The scanning beam is of very low current, about 10^{-12}A as compared with 10^{-7}A for the electron microprobe analyser, and extremely small cross section, thus the total level of X-ray excitation is low. Nevertheless the energy spectrum can be analysed and recorded and analysis is rapidly carried out in order to determine the elements lying in or near the electrode surface.

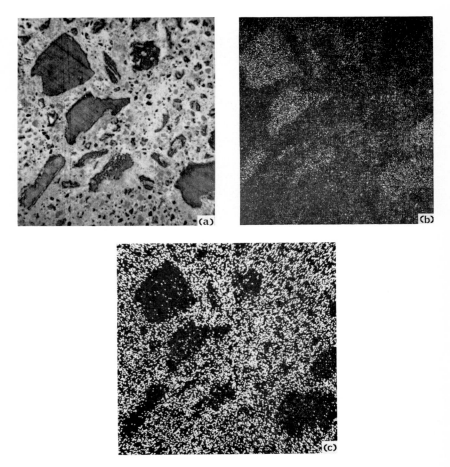

FIG. 4.14. Artifacts in the preparation of a lead electrode caused in polishing with a SiC. (a) Scanning electron micrograph, × 360; (b) set for Si excitation; (c) scanned for lead.

Three methods are available; the first by using the stationary beam located on a point chosen from the displayed image of the surface in order to select areas of chemical interest; the second to scan the surface in order to give an overall analysis; finally to display the distribution of elements by recording, by beam modulation, the location of the material of interest in the scan of the surface at a chosen magnification. Elements of as low an atomic weight as fluorine, even oxygen under certain conditions, can be detected and, it will be clear, no question of crystallinity of the material as with electron diffraction, arises. Three simple examples are given, to illustrate the technique, from routine use in the authors' laboratories. It was suspected, in kinetic studies of the reduction of chromate ions on carbon substrates, that a chromium complex was adsorbed on the electrode. The distribution of the element is displayed in Fig. 4.12(a) at a magnification of $\times 2600$ the electrode surface in Fig. 4.12(b).

Fig. 4.13(a) shows a scanning electron microscope picture of a catalysed carbon electrode with the deposit partly detached from the nickel gauze substrate; Fig. 4.13(b) shows a scanned picture with the detector set for the nickel K_α radiation; Fig. 13(c) shows the silver catalysts distribution.

The final example is selected from information observed in electrode preparation. During the polishing of a lead electrode it was suspected that SiC polishing material was being partially embedded in the lead. Figure 4.14(a) is a $\times 360$ micrograph of the surface; Fig. 4.14(b) is set for Si excitation at the same magnification and Fig. 4.14(c) scanned for lead.

ELECTRON PROBE ANALYSIS AND THE ANALYTICAL ELECTRON MICROSCOPE

The facility for the determination of elements in surface deposits, described above as one of the modes of operation of the scanning electron microscope exists also of course in a better known instrument, the X-ray microanalyser. The area under examination, however, is much larger and is only identifiable optically or by a low resolution scanning image. The scanning electron beam generates the characteristic X-rays of the specimen which are examined usually by X-ray spectrometers rather than energy dispersion measuring the intensities of the characteristic lines. A special resolution of about 1μm is obtainable and an impurity content of about 10^{19} atoms cm^{-3} may be detected. Very recently a much more refined instrument has become available for transmission specimens only combining high resolution electron microscopy with associated X-ray microanalysis and selected area diffraction facilities from areas of as little as 0·2μm diameter. The only available instrument, at the moment, is one developed by A.E.I. Scientific Apparatus Division in collaboration with the Tube Investments Research Laboratories and is designated EMMA.

References

Developments in instrumentation are rapid. Recourse to advice from instrument manufacturers is essential.

1. Bewick, A., Bewick, A., Fleischmann, M. and Liler, M. *Electrochim. Acta* **1,** 83 (1959).
2. Armstrong, R. D., Race, W. P. and Thirsk, H. R. *Electrochim. Acta,* **13,** 215 (1968).
3. (a) Technique of Organic Chemistry, Vol. III "Organic Solvents", (Eds., E. Weissburger, E. S .Priskauer, J. A. Riddick and E. E. Toops) Second Edition, Interscience, New York (1955).
   (b) Bard, A. J. "Electroanalytical Chemistry", Vol. III, Dekker, New York (1969).
4. Commercial sources for instrumentation, *In* "Servomet Spark Machine" Metals Research Ltd.,
5. (a) Piontelli, R. "Philadelphia Symposium on Electrode Processes", Electrochem. Soc., John Wiley, New York (1961).
   (b) Bouse, U. teKaat, E. and Kaplan, E. *J. Sci. Instrum.* **42,** 631 (1965).
   (c) McTegart, W. J. "The Electrolytic and Chemical Polishing of Metals", 2nd. edn., Pergamon Press, London (1959).
6. Bockris, J. O'M. and Damjanovic, A. "Modern Aspects of Electrochemistry", Vol. III, p. 224, Butterworths, London (1964).
7. (a) Pliskin, W. A. *In* "Progress in Analytical Chemistry". Vol. 2,(Eds., E. M. Murt and W. G. Guldner), Plenum Press, New York, (1969).
   (b) Wranglen, G. *Acta. Chem. Scand.* **12,** 1543 (1958); **11,** 1143 (1958).
   (c) N. Ibl. C.I.T.C.E. Proc, 7th Meeting. 112 (1952).
   (d) Wilke, C. R., Tobias, C. W. and Eisenberg. M. *J. Electrochem. Soc.* **100,** 513 (1953).
   (e) Wilke, C. R., Tobias, C. W. and Eisenberg, M. *Chem. Eng. Progr.* **49,** 663 (1963).
8. Briggs, G. W. D., private communication, Electrochemical Laboratories, University of Newcastle upon Tyne.
9. (a) Nomarski, G. *J. Phys. Radium.* **16,** 9, (1955).
   (b) Nomarski, G. and Weill, A. R. *Rev. de Metallurgie,* **52,** 121, (1955).
   (c) Allen, R. D., David, G. B. and Nomarski, G. *Z. wiss. Mikrosk. mikrosk. Tech.* **69,** 194, (1969).
10. Bockris, J. O'M. and Razummey, G. A. "Fundemental Aspects of Electrocrystallization", Plenum Press, New York (1967).
11. "Optical Studies of Adsorbed Layers at Interfaces", Faraday Society Symposium, December (1970).
12. Drude, P. *Ann. Phys. Chem.* **36,** 532, 865 (1889); **39,** 481 (1890).
13. Winterbottom, A. B. "Optical Studies of Metal Surfaces", The Royal Norwegian Scientific Society, Report No. 1, F. Brim, Trondheim, Norway (1955).
14. McCrackin, F. L., Passaglia, E., Strongberg, R. R. and Steinberg, H. L. *J. Res. Nat. Bur. Std.,* **67A,** 363 (1963).
15. "Ellipsometry in the Measurement of Surfaces and Thin Films", Symposium Proceedings, National Bureau of Standards, Washington (1963).
16. (a) Muller, R. H. and Mowat, J. R. "Analysis of Elliptically Polarised Light", University of California Radiation Laboratory, (1965).
    (b) Mowat, J. R. and Muller, R. H. "Reflection of Polarised Light from Absorbing Media", University of California Radiation Laboratory, (1966).

(c) Mowat, J. R. and Muller, R. H. "Reflection of Polarised Light from Film Covered Surfaces", University of Calfornia laboratory (1967).

(d) Muller, R. H. "Definitions and Conventions in Ellipsometry", University of California Radiation Laboratory (1968).

17. Bockris, J. O'M., Devanathan, A. V. and Reddy, A. K. N. *Proc. Roy. Soc.* **A279**, 327 (1964).

18. Bockris, J. O'M., Reddy, A. K. N. and Rao, B. Symposium on Electrode Processes, p. 281, The Electrochemical Society Spring Meeting, Cleveland (1966).

19. Hayfield, P. C. S. and White, G. W. T. Symposium proceedings, p. 157. National Bureau of Standards, Washington (1963).

20. McIntyre, J. D. E. *J. electrochem. Soc.* **116**, 1400 (1969).

21. Ying-Chech Chiu and Genshaw, M. A. *J. Phys. Chem.* **51**, 3148 (1969).

22. Gilman, S. *Electrochim. Acta* **9**, 1025 (1964).

23. Balashova, N. A. and Kagarinov, V. E. *In* "Electroanalytical Chemistry", Vol. 3, p. 125. (Ed., A. Bard) Dekker, New York (1970).

24. Feinlieb, J. *Phys. Rev. Letters,* **16**, 1200 (1966).

25. (a) Prostak, A. and Hansen, W. N. *Phys. Rev.* **160**, 600 (1967).
(b) Hansen, W. N. and Prostak, A. *Phys. Rev.* **174**, 500 (1968).
(c) Hansen, W. N. *Surface Science,* **16**, 205 (1969).

26. Hansen, W. N. *J. Opt. Soc. Amer.* **58**, 380 (1968).

27. Koch, D. F. A. *Nature,* **202**, 3871 (1964).

28. Koch, D. F. A. and Scaife, D. E. *J. electrochem. Soc.* **113**, 302 (1966).

29. McIntyre, J. D. E. and Kolbe, D. M. Faraday Society Symposium (1970).

30. Takamura, T., Takamura, W., Niffe, W. and Yeager, E. *J. electrochem. Soc.,* **117**, 626 (1970).

31. (a) Srinivasan, V. R. and Kuwana, T. *J. Phys. Chem.* **72**, 1144 (1968).
(b) Winograd, N. and Kuwana, T. *Electroanal. Chem. Interfac. Electrochem.* **23**, 333, (1969).

32. (a) Mark, H., Winstrom, L., Mattson, J, and Pons, B. *Anal. Chem.* **39**, 685 (1967).
(b) McIntyre, J. D. E. and Kolbe, D. M. Faraday Society Symposium (1970).

33. Murray, W. Heineman, W., Dom, G. W. O. *Anal. Chem.* **39**, 166 (1967).

34. Kuwana, T., Darlington, R. K. Leidy, D. W. *Anal. Chem.* **36**, 2023 (1964).

35. Hansen, W. H., Osteryoung, R. A., Kuwana, T. *Anal. Chem.* **38**, 1810 (1966).

36. Harrick, N. J. "Internal Reflections Spectrocopy". Interscience, New York (1967).

37. (a) Low, M. J. D. *J. Chem. Ed.* **41**, A97 (1969); **47**, A163, A255 (1970).
(b) Vanesse, G. A. and Sakai, H. *In* "Progress in Optics", Vol. 6, (Ed., E. Wolf), John Wiley, New York, (1967).
(c) Mertz, L. "Transformations in Optics", John Wiley, New York (1965).

38. Briggs, G. W. D. *Electrochim Acta.* **I**, 297 (1959).

39. Burbank, J. and Wales, C. P. *J. Electrochem. Soc.,* **111**, 1002 (1964).

40. "Electron Microscopy of Thin Crystals", P. B. Hush, A. Howie, R. B. Nicholson D. W. Pashley, M. J. Whelan, Butterworths, London (1965).

41. Unpublished technique. Electrochemistry Laboratories, University of Newcastle upon Tyne. Available on request.

42. Thornton, P. R. "Scanning Electron Microscopy", Chapman and Hall Ltd., London (1968).

43. Oatley, C. W., Nixon, W. C. and Pease, R. F. W. "Advances in Electronics and Electron Physics", Vol. 21, 181–247 (1965).
44. (a) Proceedings of the First Scanning Electron Microscopy Symposium, IIT Research Institute, Chicago (1968).
    (b) Second Annual S.E.M. Symposium, IIT. R.I. (1969).
    (c) Third Annual S.E.M. Symposium, IIT. R.I. (1970).
45. MacDonald, N. C. *Appl. Phys. Letters,* **16,** 76 (1969).
46. (a) Reuter, W. *Surf. Sci.* **25,** 80 (1971).
    (b) National Bureau of Standards Special Publication 298 (198).

Author Index

The numbers in brackets are the reference numbers and those in italic refer to the Reference pages where the references are listed in full. Absence of page number indicates a general reference.

163

Subject Index